IJPHM

International Journal of Prognostics and Health Management

The International Journal of Prognostics and Health Management (IJPHM) is the premier online journal related to multidisciplinary research on Prognostics, Diagnostics, and System Health Management. IJPHM is the archival journal of the Prognostics and Health Management (PHM) Society. It exists to serve the following objectives:

- To provide a focal point for dissemination of peer-reviewed PHM knowledge.
- To promote multidisciplinary collaboration in PHM education and research.
- To encourage and assure establishment of professional standards for the practice of PHM.
- To improve the professional and academic standing of all those engaged in the practice of PHM.
- To encourage governmental and industrial support for research and educational programs that will improve the PHM process and practice.

The Journal supports these goals by providing a venue for archival publication of peer-reviewed results from research and development in the area of PHM. We define PHM as a system engineering discipline focused on assessing the current status and well as predicting the future condition of a component and/or system of components. PHM is broader than any single field of engineering: it draws from electrical, electronics, mechanical, civil, and chemical engineering, computer and materials science, reliability, test and measurement, artificial intelligence, physics, and economics. IJPHM seeks to publish multidisciplinary articles from industry, academia, and government in diverse application areas such as energy, aerospace, transportation, automotive, and industrial automation. IJPHM is dedicated to all aspects of PHM: technical, management, economic, and social.

Editor-in-Chief

Neil Eklund
GE Global Research., NY
editor@ijphm.org

Associate Editors

Kai Goebel
NASA Ames Research Ctr., USA
kai.goebel@nasa.gov

Karl M. Reichard
Pennsylvania State Univ., USA
kmr5@arl.psu.edu

Irem Tumer
Oregon State Univ., USA
irem.tumer@oregonstate.edu

Serdar Uckun
Palo Alto Research Ctr.., USA
uckun@phmsociety.org

Editorial Board

SHERIF ABDELWAHED
Mississippi State Univ., USA
sherif@ece.msstate.edu

ERIC BECHHOEFER
Green Power Monitoring Systems,
eric@gpms-vt.com

JEFF BIRD
TECnos, Canada
jbird@magma.ca

GAUTAM BISWAS
Vanderbilt University, USA
gautam.biswas@vanderbilt.edu

LEONARD BOND
Iowa State University, USA
bondlj@iastate.edu

KUAN-JUNG CHUNG
Natl Changhua Univ.of Edu., TW
kjchung@cc.ncue.edu.tw

IVAN COLE
CSIRO, Australia
Ivan.Cole@csiro.au

JOHAN DE KLEER
Palo Alto Research Ctr., USA
dekleer@parc.com

FERNANDO FIGUEROA
NASA Stennis Space Ctr., USA
fernando.figueroa-1@nasa.gov

TORBJÖRN FRANSSON
SAAB AB, Sweden
torbjorn.fransson@saabgroup.com

LEN GELMAN
Cranfield University, UK
l.gelman@cranfield.ac.uk

J. WESLEY HINES
University of Tennessee, USA
jhines2@utk.edu

GIOVANNI JACAZIO
Politecnico di Torino, Italy
giovanni.jacazio@polito.it

XIAO HU
GE Research, USA
hux@research.ge.com

NARESH IYER
GE, USA
iyerna@crd.ge.com

STEPHEN JOHNSON
NASA Marshall Space Flight Ctr.
stephen.b.johnson@nasa.gov

KIRBY KELLER
Boeing Co., USA
kirby.j.keller@boeing.com

SETH KESSLER
Metis Design, USA
skessler@metisdesign.com

DANIEL LAU
City University, Hong Kong
kitlau@cityu.edu.hk

SILVAIN LETOURNEAU
NRC, Canada
Sylvain.Letourneau@nrc-cnrc.gc.ca

JAY LEE
University of Cincinnati, USA
jay.lee@uc.edu

SONY MATHEW
CALCE, University of Maryland
sonym@calce.umd.edu

HAI QIU
General Electric, USA
qiu@research.ge.com

VINCENT ROUET
EADS, France
vincent.rouet@eads.net

MIKE ROEMER
Impact Technologies, USA
Mike.Roemer@impact-tek.com

GINGER SHAO
Honeywell Intl. Inc., USA
ginger.shao@Honeywell.com

JOHN SHEPPARD
Montana State University, USA
john.sheppard@cs.montana.edu

VADIM SMELYANSKIY
NASA Ames Research Ctr., USA
Vadim.N.Smelyanskiy@nasa.gov

ASHOK N. SRIVASTAVA
NASA Ames Research Ctr., USA
ashok.n.srivastava@nasa.gov

PETER STRUSS
Technical Univ Munich, Germany
struss@in.tum.de

BO SUN
Beihang University, China
sunbo@buaa.edu.cn

GEORGE VACHTSEVANOS
Georgia Inst. of Technology, USA
gjv@ece.gatech.edu

WEIZHONG YAN
GE Global Research, USA
yan@crd.ge.com

phmsociety

PHM Society Officers

President

ANDREW HESS
Hess Associates

Vice President

JEFF BIRD
TECnos

Treasurer

KARL REICHARD
Pennsylvania State University

Secretary

PETER SANDBORN
University of Maryland

PHM Board Members

NEIL EKLUND
GE Global Research, USA

KAI GOEBEL
NASA Ames Research Center, USA

IAN JENNIONS
Cranfield University, UK

J. WESLEY HINES
University of Tennessee Knoxville, USA

JOHN MADSEN
Northrop Grumman, USA

DINKAR MYLARASWAMY
Honeywell, USA

MICHAEL ROEMER
Impact Technologies, USA

ABHINAV SAXENA
SGT, NASA Ames Research Center, USA

SERDAR UCKUN
CyDesign Labs, USA

GEORGE VACHTSEVANOS
Georgia Institute of Technology, USA

TIMOTHY WILMERING
Boeing Research and Technology, USA

IJPHM

International Journal of
Prognostics and Health Management

2010   Vol. 1 Issue 1

Table of Contents

Full Papers

Author Index

http://phmsociety.org
Free and open access to full text papers worldwide.

PROGNOSTICS AND HEALTH MANAGEMENT SOCIETY

Editorial

PHM SOCIETY established International Journal of Prognostics and Health Management (IJPHM) in 2009 to facilitate archival publication of peer-reviewed results from research and development in the area of PHM. As a journal solely dedicated to the emerging field of PHM IJPHM is the first of its kind and has been a focal point for dissemination of peer-reviewed PHM knowledge. While for the first few years the journal maintained only an online presence, the printed volumes will now be available and can be obtained upon request. PHM is broader than any single field of engineering: it draws from electrical, electronics, mechanical, civil, and chemical engineering, computer and materials science, reliability, test and measurement, artificial intelligence, physics, and economics. IJPHM publishes multidisciplinary articles from industry, academia, and government in diverse application areas such as energy, aerospace, transportation, automotive, and industrial automation. IJPHM is dedicated to all aspects of PHM: technical, management, economic, and social. In addition to regular periodic volumes IJPHM also publishes special issues with quality papers dedicated to focused topics.

The first IJPHM volume came out in 2010 with three research papers that discussed the key issue of PHM performance that is still relevant to the maturing field of PHM. Business models for PHM are still being debated and established to clearly identify the value proposition and realize commercial (or safety) value of the health forecasts. While the methods and technology have greatly improved in the last five years, some fundamental questions still remain. These three papers highlighted one of such issues and started a healthy discussion towards setting up standardized methods to implement and evaluate PHM solutions.

The first two papers dealt with metrics for evaluating the performance of PHM methods while the third one highlighted the fact that PHM has many stakeholders and that these metrics must be designed keeping user objectives of these stakeholders in mind. The paper by Saxena, Celaya, Saha, Saha, and Goebel was the first of its kind that surveyed offline prognostic metrics that incorporate probabilistic uncertainty estimates from prognostic algorithms. The paper attempted to standardize the various metrics under a unified prognostic framework and provides guidelines to assist researchers in using the metrics. The debate started by this paper is still current and pushes the state-of-the-art in prognostic performance evaluation. The second paper by Feldman, Kurtoglu, Narasimhan, Poll, Garcia, de Kleer, Kuhn, and Gemund presented a general framework, called DXF, to compare and evaluate performance of diagnostic algorithms/approaches. Using a set of standard metrics and a benchmark dataset thirteen different algorithms were evaluated. These evaluations highlight different aspects of diagnostic performance and, therefore, suggest directions for improvements as needed for each of these algorithms. Above all a method is presented to standardize performance evaluation in a systematic framework. The final paper by Wheeler, Kurtoglu, and Poll highlighted a very important aspect that goodness of performance is defined by stakeholders' user objectives and expectations. Therefore, these metrics must map to objectives that are relevant to users. A comprehensive review of cases from industry and military aerospace applications is presented to identify gaps in existing metrics towards these goals.

I am confident that this first issue of IJPHM will prove to be an indispensable reference for researchers in PHM. The PHM Society is proud to make these papers available to the global community through open access.

ABHINAV SAXENA, *Editor*
Intelligent System Division
NASA Ames Research Center
Moffett Field, CA 94035 USA

Dr. Abhinav Saxena is a Research Scientist with SGT Inc. at the Prognostics Center of Excellence of NASA Ames Research Center, Moffett Field CA. His research focus lies in developing and evaluating prognostic algorithms for engineering systems using soft computing techniques. He has co-authored more than seventy technical papers including several book chapters on topics related to PHM. He is also a member of the SAE's HM-1 committee on Integrated Vehicle Health Management Systems and IEEE working group for standards on prognostics. Dr. Saxena has been serving as editor-in-chief of the International Journal of PHM since 2011 and has led technical program committees in several PHM conferences. He has a PhD in Electrical and Computer Engineering from Georgia Institute of Technology, Atlanta. He is SGT Technical Fellow for Prognostics, has been a GM manufacturing scholar and is also a member of several professional societies including PHM Society, SAE, IEEE, AIAA, and ASME.

Welcome to the International Journal of Prognostics and Health Management (Communication)

Neil H. W. Eklund[1]

[1]General Electric Global Research, One Research Circle, Niskayuna, NY 12309, USA
editor@ijPHM.org

1. INTRODUCTION TO ijPHM

The Prognostics and Health Management Society (PHM Society) is a non-profit organization dedicated to the advancement of PHM as an engineering discipline. One of the fundamental principles of the Society is to provide centralized, timely, free and unrestricted access to knowledge about PHM research and applications.

The flagship publication of the PHM Society is the open online journal, International Journal of Prognostics and Health Management (ijPHM), publishing multidisciplinary research on Prognostics, Diagnostics, and System Health Management. In contrast to most other technical organizations, PHM Society has adopted a Creative Commons license policy towards its archival journal which allows authors to retain copyright while allowing the Society to distribute their work broadly through modern media. Papers published in ijPHM are available for download to everyone, everywhere, without restriction, and at no cost.

The focus of ijPHM is to promote the exchange of innovative ideas and to advance PHM as a scientific discipline. The editorial team at ijPHM is dedicated to providing a very short publication cycle – papers are reviewed within eight weeks of initial submission, and articles are published as soon as an acceptably revised version is received. This pace is much faster than what is possible with traditional print media, allowing results from PHM research and application to be broadly disseminated while they are most relevant, promoting rapid progress in the field. However, this fast pace does not diminish the level of scientific rigor required for publication – each article will be subjected to peer review by at least three independent scientists with established track records in the field.

Because ijPHM is an online publication, it is free of the limitations of print media. The length of regular journal articles is unrestricted, placing no arbitrary limit on the number of pages necessary to describe and interpret the work clearly in the context of other research. The appropriate use of color in figures and color photographs are encouraged. Raw or processed experimental data and video that elegantly illustrates specific, relevant scientific aspects of the paper may be attached to the publication. Papers may also include links to web sites, other publications, and to other sections of the same document. Looking ahead, we are also considering unconventional ideas such as "living articles" that can be revised and expanded over time, on-demand printing of customized volumes, and allowing public discussion threads for published papers.

2. INFORMATION FOR AUTHORS

ijPHM publishes scientific papers dealing with all aspects of prognostics, diagnostics, and system health management of complex engineered systems. We accept high quality articles focused on assessing the current status and predicting the future condition of an engineered component and/or system of components. Such articles may come from a variety of disciplines, including electrical, electronics, mechanical, civil, and chemical engineering, computer and materials science, reliability, test and measurement, artificial intelligence, physics, and economics. Contributions to ijPHM must report original research and will be subjected to review by referees at the discretion of the Editor. ijPHM considers only manuscripts that have not been published elsewhere (including at conferences), and that are not under consideration for publication or in press elsewhere. Moreover, it is the responsibility of the author to ensure that any data or information submitted complies with the export-control regulations of the author's home country, e.g., International Traffic in Arms Regulations (ITAR) in the United States.

This is an open-access article distributed under the terms of the Creative Commons Attribution 3.0 United States License, which permits unrestricted use, distribution, and reproduction in any medium, provided the original author and source are credited.

Submitted 10/2009; published 10/2009.

ijPHM publishes full-length regular papers, technical briefs, communications, and survey papers. **Full-length regular papers** should describe new and carefully confirmed findings. The experimental methods used, results obtained, analysis, and conclusions must be presented clearly and objectively, such that they might be replicated by another researcher. A full paper should be long enough to describe and interpret the work clearly, placing it in the context of other research.

Technical briefs typically describe a single result, experiment, or technique of general interest for which a short treatment is appropriate. Any experimental methods used, results obtained, analysis, and conclusions must be presented clearly and objectively, such that they might be replicated by another researcher. A technical brief should be long enough to describe and interpret the work clearly, although not necessarily in the context of other research.

Communications are a separate class of short manuscripts that are subject to an expedited review process. Appropriate items include (but are not limited to) rebuttals and/or counterexamples of previously published papers. A communication is suitable for highlighting the results of newly-completed projects or providing details of new models or hypotheses, innovative methods, techniques or apparatus. The style of main sections need not conform to that of full-length papers. Short communications are 2 to 4 printed pages in length. The Editors will review these submissions internally, and request outside review when appropriate.

Survey papers covering emerging research topics in PHM are also published, and unsolicited manuscripts of a tutorial or review nature are welcome. However, prospective authors of survey papers should contact the Editor-in-Chief in advance in order to assess the possible appeal of the topic for publication in ijPHM. Papers describing specific current applications are encouraged, provided that the designs represent the best current practice, detailed characteristics and performance are included, the application is discussed in the context of the state-of-the-art in the field, and it is of general interest.

Prospective authors should note that ijPHM is focused on promoting the exchange of innovative ideas and advancing PHM as a scientific discipline. Poorly documented papers or papers using "proprietary" techniques may be rejected without a comprehensive review. Moreover, ijPHM should not be seen as advertising space! Excessive "branding" within a paper is also cause for rejection; e.g., "The CompanyBrand™ team used the magical CompanyBrand™ preprocessing technique to prepare the data to extract the amazing CompanyBrand™-proprietary features (which we can't tell you about)." Generally, no specific mention of a particular business (except where necessary to specify a piece of equipment) is permitted beyond the author's affiliation listed beneath their name, or sponsoring organizations given credit in the acknowledgements section.

ijPHM provides a very short publication cycle. Papers are reviewed initially to screen for obvious fatal defects – e.g., unacceptable English usage, excessive branding, clearly insufficient context (cited literature), etc. Acceptable papers are then sent to one of the members of the Editorial Board to be assigned reviewers. All full-length papers are reviewed by a minimum of three qualified reviewers. Specific peer-review feedback is provided to the author within eight weeks of initial submission. If the paper is approved for publication, necessary revisions should be made and submitted within four weeks.

3. CALL FOR PAPERS

The Prognostics and Health Management Society invites you to submit scientific papers of the highest quality dealing with all aspects of prognostics, diagnostics, and system health management of complex engineered systems to ijPHM. Areas of interest include but are not limited to:

- PHM system design and engineering
- Physics of failure
- Software health management
- Structural health management
- PHM for electronics
- Health management for renewable energies
- Diagnosis methods
- Data-driven prognostics
- Model-based prognostics
- Standards and methodologies
- Fault-adaptive controls
- Technology maturation
- Return-on-investment analysis
- Deployed applications and success stories

We welcome both papers that focus on fundamental research and application-oriented papers from diverse application areas such as energy, aerospace, transportation, automotive, and industrial automation.

Metrics for Offline Evaluation of Prognostic Performance

Abhinav Saxena[1], Jose Celaya[1], Bhaskar Saha[2],
Sankalita Saha[2], and Kai Goebel[3]

[1]SGT Inc., NASA Ames Research Center, Intelligent Systems Division, Moffett Field, CA 94035, USA
abhinav.saxena@nasa.gov
jose.r.celaya@nasa.gov

[2]MCT Inc., NASA Ames Research Center, Intelligent Systems Division, MS 269-4, Moffett Field, CA 94035, USA
bhaskar.saha@nasa.gov
sankalita.saha-1@nasa.gov

[3]NASA Ames Research Center, Intelligent Systems Division, MS 269-4, Moffett Field, CA 94035, USA
kai.goebel@nasa.gov

ABSTRACT

Prognostic performance evaluation has gained significant attention in the past few years. Currently, prognostics concepts lack standard definitions and suffer from ambiguous and inconsistent interpretations. This lack of standards is in part due to the varied end-user requirements for different applications, time scales, available information, domain dynamics, etc. to name a few. The research community has used a variety of metrics largely based on convenience and their respective requirements. Very little attention has been focused on establishing a standardized approach to compare different efforts. This paper presents several new evaluation metrics tailored for prognostics that were recently introduced and were shown to effectively evaluate various algorithms as compared to other conventional metrics. Specifically, this paper presents a detailed discussion on how these metrics should be interpreted and used. These metrics have the capability of incorporating probabilistic uncertainty estimates from prognostic algorithms. In addition to quantitative assessment they also offer a comprehensive visual perspective that can be used in designing the prognostic system. Several methods are suggested to customize these metrics for different applications. Guidelines are provided to help choose one method over another based on distribution characteristics. Various issues faced by prognostics and its performance evaluation are discussed followed by a formal notational framework to help standardize subsequent developments.

This is an open-access article distributed under the terms of the Creative Commons Attribution 3.0 United States License, which permits unrestricted use, distribution, and reproduction in any medium, provided the original author and source are credited.

Submitted 1/2010; published 4/2010.

1. INTRODUCTION

In the systems health management context, prognostics can be defined as predicting the Remaining Useful Life (RUL) of a system from the inception of a fault based on a continuous health assessment made from direct or indirect observations from the ailing system. By definition prognostics aims to avoid catastrophic eventualities in critical systems through advance warnings. However, it is challenged by inherent uncertainties involved with future operating loads and environment in addition to common sources of errors like model inaccuracies, data noise, and observer faults among others. This imposes a strict validation requirement on prognostics methods to be proven and established though a rigorous performance evaluation before they can be certified for critical applications.

Prognostics can be considered an emerging research field. Prognostic Health Management (PHM) has in most respects been accepted by the engineered systems community in general, and by the aerospace industry in particular, as a promising avenue for managing the safety and cost of complex systems. However, for this engineering field to mature, it must make a convincing business case to the operational decision makers. So far, in the early stages, focus has been on developing prognostic methods themselves and very little has been done to define methods to allow comparison of different algorithms. In two surveys on methods for prognostics, one on data-driven methods (Schwabacher, 2005) and one on artificial-intelligence-based methods (Schwabacher & Goebel, 2007), it can be seen that there is a lack of standardized methodology for performance evaluation and in many cases performance evaluation is not even formally addressed. Even the current ISO standard by International Organization for Standards (ISO, 2004) for prognostics in condition monitoring and diagnostics of machines lacks a firm definition of any such methods. A dedicated effort to develop methods and metrics to evaluate prognostic algorithms is needed.

Metrics can create a standardized language with which technology developers and users can communicate their findings and compare results. This aids in the dissemination of scientific information as well as decision making. Metrics could also be viewed as a feedback tool to close the loop on research and development by using them as objective functions to be optimized as appropriate by the research effort.

Recently there has been a significant push towards crafting suitable metrics to evaluate prognostic performance. Researchers from government, academia, and industry are working closely to arrive at useful performance measures. With these objectives in mind a set of metrics have been developed and proposed to the PHM community in the past couple years (Saxena, Celaya, Saha, Saha, & Goebel, 2009b). These metrics primarily address algorithmic performance evaluation for prognostics applications but also have provisions to link performance to higher level objectives through performance parameters. Based on experience gained from a variety of prognostic applications these metrics were further refined. The current set of prognostics metrics aim to tackle offline performance evaluation methods for applications where run-to-failure data are available and true End-of-Life (EoL) is known *a priori*. They are particularly useful for the algorithm development phase where feedback from the metrics can be used to fine-tune prognostic algorithms. These metrics are continuously evolving and efforts are underway towards designing on-line performance metrics. This will help associate a sufficient degree of confidence to the algorithms and allow their application in real *in-situ* environments.

1.1 Main Goals of the Paper

This paper presents a discussion on prognostics metrics that were developed in NASA's Integrated Vehicle Health Management (IVHM) project under the Aviation Safety program (NASA, 2009). The paper aims to make contribution towards providing the reader with a better understanding of:

- the need for separate class of prognostic performance metrics
- difference in user objectives and corresponding needs from a performance evaluation view point
- what can or cannot be borrowed from other forecasting related disciplines
- issues and challenges in prognostics and prognostic performance evaluation
- key prognostic concepts and a formal definition of a prognostic framework

- new performance evaluation metrics, their application and interpretation of results
- research issues and other practical aspects that need to be addressed for successful deployment of prognostics

1.2 Paper Organization

Section 2 motivates the development of prognostic metrics. A comprehensive literature review of performance assessment for prediction/forecasting applications is presented in section 3. This section also categorizes prognostic applications in several classes and identifies the differences from other forecasting disciplines. Key aspects for prognostic performance evaluation are discussed in Section 4. Technical development of new performance metrics and a mathematical framework for the prognostics problem are then presented in detail in Section 5. Section 6 follows with a brief case study as an example for application of these metrics. The paper ends with future work proposals and concluding discussions in Sections 7 and 8 respectively.

2. MOTIVATION

This research is motivated by two-fold benefits of establishing standard methods for performance assessment (see Figure 1). One, it will help create a foundation for assessing and comparing performance of various prognostics methods and approaches as far as low level algorithm development is concerned. Two, from a top-down perspective, it will help generate specifications for requirements that are imposed by cost-benefit and risk constraints at different system lifecycle stages in order to ensure safety, availability, and reliability. In this paper we discuss these metrics primarily in the context of the first benefit and only a brief discussion is provided on requirements specification.

2.1 Prognostic Performance Evaluation

Most of the published work in the field of prognostics has been exploratory in nature, such as proof-of-concepts or one-off applications. A lack of standardized guidelines has led researchers to use common accuracy and precision based metrics, mostly borrowed from the diagnostics domain. In some cases these are modified on an ad hoc basis to suit specific applications. This makes it rather difficult to compare various efforts and choose a winning candidate from several algorithms, especially for safety critical applications. Research efforts are focusing on developing algorithms that can provide a RUL estimate, generate a confidence bound around the predictions, and be easily integrated with existing diagnostic systems. A key step in successful deployment of a PHM system is prognosis certification. Since prognostics is still considered relatively immature (as compared to diagnostics), more focus so far has been on developing prognostic methods rather than evaluating and comparing their performances. Consequently, there is a need for dedicated attention towards developing standard methods to evaluate prognostic performance from a viewpoint of how post prognostic reasoning will be integrated into the health management decision making process.

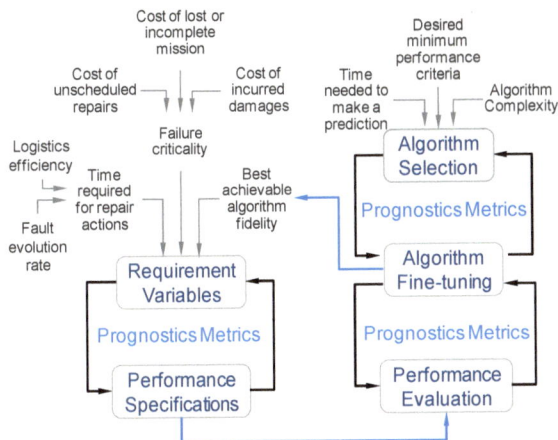

Figure 1: Prognostics metrics facilitate performance evaluation and also help in requirements specification.

2.2 Prognostic Requirements Specification

Technology Readiness Level (TRL) for the current prognostics technology is considered low. This can be attributed to several factors lacking today such as

- assessment of *prognosability* of a system,
- concrete Uncertainty Representation and Management (URM) approaches,
- stringent Validation and Verification (V&V) methods for prognostics
- understanding of how to incorporate risk and reliability concepts for prognostics in decision making

Managers of critical systems/applications have consequently struggled while defining concrete prognostic performance specifications. In most cases, performance requirements are either derived from prior experiences like diagnostics in Condition Based Maintenance (CBM) or are very loosely specified. This calls for a set of performance metrics that not only encompass key aspects of predicting into the future but also accommodate notions from practical aspects such as logistics, safety, reliability, mission criticality, economic viability, etc. The key concept that ties all these notions in a prognostic framework is of *performance tracking* as time evolves while various trade-offs continuously arise in a dynamic situation. The prognostics metrics presented in this paper are

designed with intentions to capture these salient features. Methodology for generating requirements specification is beyond the scope of this paper and only a brief discussion explaining these ideas is provided in the subsequent sections.

3. LITERATURE REVIEW

As research activities gain momentum in the area of PHM, efforts are underway to standardize prognostics research(Uckun, Goebel, & Lucas, 2008). Several studies provide a detailed overview of prognostics along with its distinction from detection and diagnosis (Engel, 2008; Engel, Gilmartin, Bongort, & Hess, 2000). The importance of uncertainty management and the various other challenges in determining remaining useful life are well presented. Understanding the challenges in prognostics research is an important first step in standardizing the evaluation and performance assessment. Thus, we draw on the existing literature and provide an overview of the important concepts in prognostic performance evaluation before defining the new metrics.

3.1 Prediction Performance Evaluation Methods

Prediction or forecasting applications are common in medicine, weather, nuclear, finance and economics, automotive, aerospace, and electronics. Metrics based on accuracy and precision with slight variations are most commonly used in all these fields in addition to a few metrics customized to the domain. In medicine and finance, statistical measures are heavily used exploiting the availability of large datasets. Predictions in medicine are evaluated based on hypothesis testing methodologies while in finance errors calculated based on reference prediction models are used for performance evaluation. Both of them use some form of precision and accuracy metrics such as MSE (mean squared error), SD (standard deviation), MAD (mean absolute deviation), MdAD (median absolute deviation), MAPE (mean absolute percentage error) and similar variants. Other domains like aerospace, electronics, and nuclear are relatively immature as far as fielded prognostics applications are concerned. In addition to conventional accuracy and precision measures, a significant focus has been on metrics that assess business merits such as ROI (return on investment), TV (technical value), and life cycle cost, rather than reliability based metrics like MTBF (mean time between failure) or the ratio MTBF/MTBUR (mean time between unit replacements). Notions of false positives, false negatives and ROC (receiver operator characteristics) curves have also been adapted for prognostics (Goebel & Bonissone, 2005).

3.2 Summary of the Review

Active research and the quest to find out what constitutes performance evaluation in forecasting related tasks in other domains painted a wider landscape of requirements and domain specific characteristics than initially anticipated. This naturally translated into identifying the similarities and differences in various prediction applications to determine what can or cannot be borrowed from those domains. As shown in Figure 2, a classification tree was generated that listed key characteristics of various forecasting applications and examples of domains that exhibited those (Saxena et al., 2008).

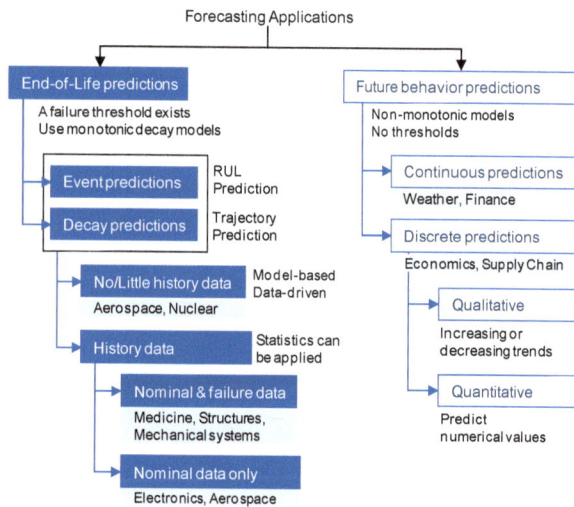

Figure 2: Categories of the forecasting applications (Saxena, et al., 2008).

Coble & Hines (2008) categorized prognostic algorithms into three categories based on type of models/information used for predictions. These types of information about operational and environmental loads are an inherent part of prognostic problems and must be used wherever available. From the survey it was identified that not only did the applications differ in nature, the metrics within domains also varied based on functionality and nature of the end use of performance data. This led to classifying the metrics based on end usage (see Table 1) and their functional characteristics (Figure 3). In a similar effort end users were classified from a health management stakeholder's point of view (Wheeler, Kurtoglu, & Poll, 2009). Their top-level user groups include Operations, Regulatory, and Engineering. It was observed that it was prognostics algorithm performance that translated into valuable information for these user groups in form or another. For instance, it can be argued that low level algorithmic performance metrics are connected to operational and regulatory branches through a requirement specification process. Therefore, further attention in this effort was focused on algorithmic performance metrics.

Table 1: Classification of prognostic metrics based on end user requirements as adapted from Saxena, et al. (2008) and Wheeler, et al. (2009).

Category	End User	Goals	Metrics
Operations	Program Manager	Assess the economic viability of prognosis technology for specific applications before it can be approved and funded.	Cost-benefit type metrics that translate prognostics performance in terms of tangible and intangible cost savings.
	Plant Manager	Resource allocation and mission planning based on available prognostic information.	Accuracy and precision based metrics that compute RUL estimates for specific Unit Under Test (UUT). Such predictions are based on degradation or damage accumulation models.
	Operator	Take appropriate action and carry out re-planning in the event of contingency during mission.	Accuracy and precision based metrics that compute RUL estimates for specific UUTs. These predictions are based on fault growth models for critical failures.
	Maintainer	Plan maintenance in advance to reduce UUT downtime and maximize availability.	Accuracy and precision based metrics that compute RUL estimates based on damage accumulation models.
Engineering	Designer	Implement the prognostic system within the constraints of user specifications. Improve performance by modifying design.	Reliability based metrics to evaluate a design and identify performance bottlenecks. Computational performance metrics to meet resource constraints.
	Researcher	Develop and Implement robust performance assessment algorithms with desired confidence levels.	Accuracy and Precision based metrics that employ uncertainty management and output probabilistic predictions in presence of uncertain conditions.
Regulatory	Policy Makers	To assess potential hazards (safety, economic, and social) and establish policies to minimize their effects.	Cost-benefit-risk measures, Accuracy and Precision based RUL measures to establish guidelines & timelines for phasing out of aging fleet and/or resource allocation for future projects.

There are different types of outputs from various prognostic algorithms. Some algorithms assess Health Index (HI) or Probability of Failure (PoF) at any given point and others carry out an assessment of RUL based on a predetermined Failure Threshold (FT) (Coble & Hines, 2008; Orsagh, Roemer, Savage, & McClintic, 2001; Saxena, et al., 2008). The ability to generate representations of uncertainty for predictions such as probability distributions, fuzzy membership functions, possibility distribution, etc., further distinguishes some algorithms from others that generate only point estimates of the predictions. This led to the conclusion that a formal prognostic framework must be devised and additional performance metrics needed to be developed to accommodate most of these scenarios in an intuitive way.

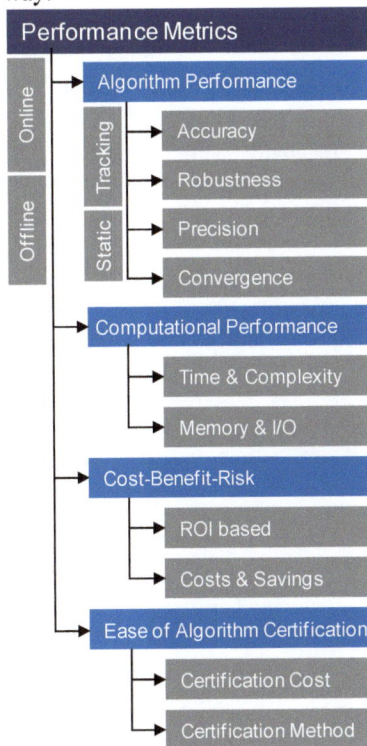

Figure 3: Functional classification of prognostics metrics (adapted from Saxena, et al. (2008)).

For further details on these classifications and examples of different applications the reader is referred to Saxena, et al. (2008).

3.3 Recent Developments in the PHM Domain

To update the survey conducted in Saxena, et al. (2008) relevant developments were tracked during the last two years. A significant push has been directed towards developing metrics that measure economic viability of prognostics. In Leao, et al. (2008) authors suggested a variety of metrics for prognostics based on commonly used diagnostic metrics. Metrics like false positives and negatives, prognostics effectiveness, coverage, ROC curve, etc. were suggested with slight modifications to their original definitions. Attention was more focused on integrating these metrics into user requirements and cost-benefit analysis. A simple tool is introduced in Drummond & Yang (2008) to evaluate a prognostic algorithm by estimating the cost savings expected from its deployment. By accounting for variable repair costs and changing failure probabilities this tool is useful for demonstrating the cost savings that prognostics can yield at the operational levels. A commercial tool to calculate the Return on Investment (ROI) for prognostics for electronics systems was developed (Feldman, Sandborn, & Jazouli, 2008). The 'returns'

that are considered could be the cost savings, profit, or cost avoidance by the use of prognostics in a system. Wheeler, et al. (2009) compiled a comprehensive set of user requirements and mapped them to performance metrics separately for diagnostics and prognostics.

For algorithm performance assessment, Wang & Lee (2009) proposed simple metrics adapted from the classification discipline and also suggested a new metric called "Algorithm Performance Profile" that tracks the performance of an algorithm using a accuracy score each time a prediction is generated. In Yang & Letourneau (2007), authors presented two new metrics for prognostics. They defined a reward function for predicting the correct time-to-failure that also took into account prediction and fault detection coverage. They also proposed a cost-benefit analysis based metric for prognostics. In some other approaches model based techniques are adopted where discrete event simulations are run and results evaluated based on different degrees of prediction error rates (Carrasco & Cassady, 2006; Pipe, 2008). These approaches are beyond the scope of the current discussion.

4. CHALLENGES IN PROGNOSTICS

There are several unsolved issues in prognostics that complicate the performance evaluation task. These complications share partial responsibility for the lack of standardized procedures. A good set of metrics should accommodate all or most of these issues but not necessarily require all of them to have been addressed together in any single application. Enumerating these issues briefly here should help understanding the discussions on metrics development later.

Acausality: Prognostics is an *acausal* problem that requires an input from future events, for instance the knowledge about operational conditions and load profiles in order to make more accurate predictions. Similarly, to accurately assess the performance (accuracy or precision) one must know the EoL to compare with the predicted EoL estimates. Where the knowledge about these quantities is rarely and completely available, some estimates can be derived based on past usage history, plan for the mission profile, and predictions for future operating and environmental conditions that are not controllable (e.g., weather conditions). This however, adds uncertainty to the overall process and makes it difficult to judiciously evaluate prognostic performance.

Run-to-Failure Data from Real Applications: Another aspect that makes this evaluation further complicated is considered the paradox of prognostics – "Not taking an action on a failure prediction involves the risk of failure and an action (e.g. system maintenance and repair), on the contrary, eliminates all chances of validating the correctness of the prediction itself". Therefore, it has been a challenging task to assess long term prognostic results. For instance, consider the following scenario where aircraft engines undergo continuous monitoring for fault conditions and scheduled maintenance for system deterioration. In the PHM context a decision about when to perform the maintenance, if not scheduled, is a rather complex one that should be based on current health condition, next flight duration, expected operational (weather) conditions, availability of spares and a maintenance opportunity, options available for alternate planning, costs, risk absorbing capacity, etc. In this situation one could arguably evaluate a prognostic result against statistical (reliability) data about the RULs from similar systems. However, in practice such data are rarely available because there are typically very few faults that were allowed to go all the way to a failure resulting perhaps in an extremely unavoidable in-flight engine shutdown or an aborted takeoff. Furthermore, once the maintenance operation has been performed two problems arise from the perspective of performance evaluation. One, there is no way to verify whether the failure prediction was indeed correct, and two, the useful life of the system has now changed and must have moved the EoL point in time from its previous estimate. Alternatively, allowing the system to fail to evaluate the prognosis would be cost and safety prohibitive.

Offline Performance Evaluation: The aforementioned considerations lead to an argument in favor of controlled run-to-failure (RtF) experiments for the algorithm development phase. While this makes it simpler for the offline performance evaluation some issues still remain. First, it is difficult to extend the results of offline setup to a real-time scenario. Second, often in an RtF experiment the setup needs frequent disassemblies to gather ground truth data. This assembly-disassembly process creates variations in the system performance and the EoL point shifts from what it may have been in the beginning of the experiment. Since actual EoL is observed only at the end there is no guarantee that a prediction made based on initial part of data will be very accurate. Whereas, this does not necessarily mean that prognostic algorithm is poorly trained, it is difficult to confirm otherwise. Therefore, one must be careful while interpreting the performance assessment results. Third, even controlled RtF experiments can be very expensive and time consuming, in particular if one seeks to conduct statistically significant number of experiments for all components and fault modes.

There is no simple answer to tackle these issues. However, using reasonable assumptions they can be tackled one step at a time. For instance, most prognostics algorithms make implicit assumptions of perfect knowledge about the future in a variety of ways such as following:

- operating conditions remain within expected bounds more or less throughout systems life
- any change in these conditions does not affect the life of the system significantly, or
- any controllable change (e.g., operating mode profile) is known (deterministically or probabilistically) and is used as an input to the algorithm

Although these assumptions do not hold true in most real-world situations, the science of prognostics can be advanced and later improved by making adjustments for them as new methodologies develop.

Uncertainty in Prognostics: A good prognostics system not only provides accurate and precise estimates for the RUL predictions but also specifies the level of confidence associated with such predictions. Without such information any prognostic estimate is of limited use and cannot be incorporated in mission critical applications (Uckun, et al., 2008). Uncertainties arise from various sources in a PHM system (Coppe, Haftka, Kim, & Yuan, 2009; Hastings & McManus, 2004; Orchard, Kacprzynski, Goebel, Saha, & Vachtsevanos, 2008). Some of these sources include:

- modeling uncertainties (modeling errors in both system model and fault propagation model),
- measurement uncertainties (arise from sensor noise, ability of sensor to detect and disambiguate between various fault modes, loss of information due to data preprocessing, approximations and simplifications),
- operating environment uncertainties,
- future load profile uncertainties (arising from unforeseen future and variability in usage history data),
- input data uncertainties (estimate of initial state of the system, variability in material properties, manufacturing variability), etc.

It is often very difficult to assess the levels and characteristics of uncertainties arising from each of these sources. Further, it is even more difficult to assess how these uncertainties that are introduced at different stages of the prognostic process combine and propagate through the system, which most likely has a complex non-linear dynamics. This problem worsens if the statistical properties do not follow any known parametric distributions allowing analytical solutions.

Owing to all of these challenges Uncertainty Representation and Management (URM) has become an active area of research in the field of PHM. A conscious effort in this direction is clearly evident from recent developments in prognostics (DeNeufville, 2004; Ng & Abramson, 1990; Orchard, et al., 2008; Sankararaman, Ling, Shantz, & Mahadevan, 2009;

Tang, Kacprzynski, Goebel, & Vachtsevanos, 2009). These developments must be adequately supported by suitable methods for performance evaluation that can incorporate various expressions of uncertainties in the prognostic outputs.

Although several approaches for uncertainty representation have been explored by researchers in this area, the most popular approach has been probabilistic representation. A well founded Bayesian framework has led to many analytical approaches that have shown promise (Guan, Liu, Saxena, Celaya, & Goebel, 2009; Orchard, Tang, Goebel, & Vachtsevanos, 2009; Saha & Goebel, 2009). In these cases a prediction is represented by a corresponding Probability Density Function (PDF). When it comes to performance assessment, in many cases a simplifying assumption is made about the form of distribution being Normal or any other known probability distribution. The experience from several applications, however, shows that this is hardly ever the case. Mostly these distributions are non-parametric and are represented by sampled outputs.

This paper presents prognostic performance metrics that incorporate these cases irrespective of their distribution characteristics.

5. PROGNOSTIC FRAMEWORK

First, a notational framework is developed to establish relevant context and terminology for further discussions. This section provides a list of terms and definitions that will be used to describe the prognostics problem and related concepts to develop the performance evaluation framework. Similar concepts have been described in the literature. They sometimes use different terms to describe different concepts. This section is intended to resolve ambiguities in interpreting these terms for the purpose of discussions in this paper. It must be noted that in the following discussions $t_x \in \Re^+$ is used to denote time expressed in absolute units e.g., hours, minutes, seconds, etc., and $x \in I^+$ is a time index to express time in relative units like operating hours, cycles, etc. It follows from the fact that realistic data systems sample from real continuous physical quantities.

Table 2: Frequently used prognostic terms and time indexes to denote important events in a prognostic process.

Prognostic Terms	
UUT	Unit Under Test – an individual system for which prognostics is being developed. Although the same methodology may be applicable for multiple systems in a fleet, life predictions are generated specific to each UUT.
PA	Prognostic Algorithm – An algorithm that tracks and predicts the growth of a fault mode with time. PA may be data driven, model-based or a hybrid.
RUL	Remaining Useful Life – amount of time left for which a UUT is usable before some corrective action is required. It can be specified in relative or absolute time units, e.g., load cycles, flight hours, minutes, etc.
FT	Failure Threshold – a limit on damage level beyond which a UUT is not usable. FT does not necessarily indicate complete failure of the system but a conservative estimate beyond which risk of complete failure exceeds tolerance limits.
RtF	Run-to-Failure – refers to a scenario where a system has been allowed to fail and corresponding observation data are collected for later analysis.

Important Time Index Definitions (Figure 4)	
t_0	Initial time when health monitoring for a UUT begins.
F	Time index when a fault of interest initiates in the UUT. This is an event that might be unobservable until the fault grows to detectable limits.
D	Time index when a fault is detected by a diagnostic system. It denotes the time instance when a prognostic routine is triggered the first time.
P	Time index when a prognostics routine makes its first prediction. Generally speaking, there is a finite delay before predictions are available once a fault is detected.
EoL	End-of-Life – time instant when a prediction crosses a FT. This is determined through RtF experiments for a specific UUT.
EoP	End-of-Prediction – time index for the last prediction before EoL is reached. this is a conceptual time index that depends on frequency of prediction and assumes predictions are updated until EoL is reached.
EoUP	End-of-Useful Predictions – time index beyond which it is futile to update a RUL prediction because no corrective action is possible in the time available before EoL.

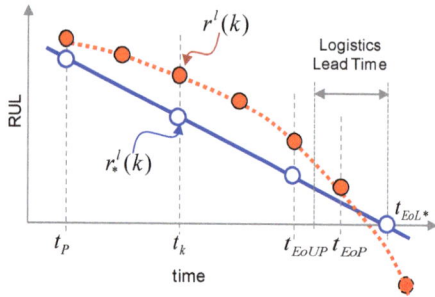

Figure 4: An illustration depicting some important prognostic time indices (definitions and concepts).

Symbols and Notations

i	time index representing time instant t_i
l	is the index for l^{th} unit under test (UUT)
p	set of all indexes when a prediction is made the first element of p is P and the last is EoP
t_{EoL}	time instant at End-of-Life (EoL)
t_{EoUP}	time for End-of-Useful-Prediction (EoUP)
t_{repair}	time taken by a reparative action for a system
t_P	time instant when the first prediction is made
t_D	time instant when a fault is detected
$f_n^l(i)$	n^{th} feature value for the lth UUT at time t_i
$c_m^l(i)$	m^{th} operational condition for the lth UUT at t_i
$r^l(i)$	predicted RUL for the lth UUT at time t_i reference to l may be omitted for a single UUT
$r_*^l(i)$	ground truth for RUL at time t_i
$\phi^l(i\,|\,j)$	Prediction for time t_i given data up to time t_j for the lth UUT. Prediction may be made in any domain, e.g., feature, health, RUL, etc.
$\underline{\Phi}^l(i)$	Trajectory of predictions $\phi^l(i\,|\,j)$ made for the lth UUT at time t_j for all times t_i s.t. $i > j$. E.g., financial and weather forecasts
$h^l(i)$	Health of system for the lth UUT at time t_i
α	accuracy modifier such that $\alpha \in [0,1]$
α^+	maximum allowable positive error
α^-	minimum allowable negative error
λ	time window modifier s.t. $t_\lambda = t_P + \lambda(t_{EoL} - t_P)$ where $\lambda \in [0,1]$
β	minimum desired probability threshold
ω	weight factor for each Gaussian component
θ	parameters of RUL distribution
$\varphi(x)$	non-parameterized probability distribution for any variable x
$\varphi_\theta(x)$	parameterized probability distribution for any variable x
$\pi[x]$	probability mass of a distribution of any variable x within α-bounds $[\alpha^-, \alpha^+]$, i.e. $\pi[x] = \sum_{\alpha^-}^{\alpha^+} \varphi(x); x \in I^+$ or $\int_{\alpha^-}^{\alpha^+} \varphi_\theta(x)dx; x \in \Re^+$
$M(i)$	a performance metric of interest at time t_i

C_M	center of mass as a measure of convergence for a metric M
x_c, y_c	x and y coordinates for center of mass (C_M)

Assumptions for the Framework

- Prognostics is condition based health assessment that includes detection of failure precursors from sensor data, prediction of RUL by generating a current state estimate and using expected future operational conditions for a specific system.

- A suitable diagnostic algorithm correctly detects, identifies and isolates the system fault before it triggers a PA to predict evolution for that specific fault mode.

- If the information about future operational conditions is available it may be explicitly used in the predictions. Any prediction, otherwise, implicitly assumes current conditions would remain in the future and/or variations from current operating conditions do not affect the life of a system.

- RUL estimation is a prediction/ forecasting/ extrapolation process.

- Algorithms incorporate uncertainty representation and management methods to produce RUL distributions. Point estimates for RUL may be generated from these distributions through suitable methods when needed.

- RtF data are available that include sensor measurements, operating condition information, and EoL ground truth.

- A definition of failure threshold is available that determines the EoL for a system beyond which the system is not recommended for further use.

- In the absence of true EoL (determined experimentally) statistical (reliability) data such as MTTF (Mean Time to Failure) or MTBF (Mean Time Between Failures) may be used to define EoL with appropriate caution.

In a generic scenario a PA is triggered by an independent diagnostic algorithm whenever it detects a fault in the system with high certainty. PA may take some time to gather more data and tune itself before it starts predicting the growth of that fault. Based on a user defined FT the PA determines where the fault is expected to cross the FT and EoL of the system is reached. An estimate of RUL is generated by computing the difference between estimated EoL and the current time. As time progresses more measurement data become available that are used to make another prediction and the estimates of EoL and RUL are correspondingly updated. This process continues until one of the following happens:

- the system is taken down for maintenance.
- EoUP is reached and any further predictions may not be useful for failure avoidance operations.
- the system has failed (unexpectedly).
- the case where problem symptoms have disappeared (can occur if there were false alarms, intermittent fault, etc.).

Definitions

Time Index: In a prognostics application time can be discrete or continuous. A time index i will be used instead of the actual time, e.g., $i=10$ means t_{10}. This takes care of cases where sampling time is not uniform. Furthermore, time indexes are invariant to time-scales.

Time of Detection of Fault: Let D be the time index for time (t_D) at which the diagnostic or fault detection algorithm detected the fault. This process will trigger the prognostics algorithm which should start making RUL predictions as soon as enough data has been collected, usually shortly after the fault was detected. For some applications, there may not be an explicit declaration of fault detection, e.g., applications like battery health management, where prognosis is carried out on a decay process. For such applications t_D can be considered equal to t_0 i.e., prognostics is expected to trigger as soon as enough data has been collected instead of waiting for an explicit diagnostic flag (see Figure 5).

Time to Start Prediction: Time indices for times at which a fault is detected (t_D) and when the system starts predicting (t_P) are differentiated. For certain algorithms $t_D = t_P$ but in general $t_P \geq t_D$ as PAs need some time to tune with additional fault progression data before they can start making predictions (Figure 5). Cases where a continuous data collection system is employed even before a fault is detected, sufficient data may already be available to start making predictions and hence $t_P = t_D$.

Prognostics Features: Let $f_n^l(i)$ be a feature at time index i, where $n = 1, 2, ..., N$ is the feature index, and $l = 1, 2, ..., L$ is the UUT index (an index identifying the different units under test). In prognostics, irrespective of the analysis domain, i.e., time, frequency, wavelet, etc., features take the form of time series and can be physical variables, system parameters or any other quantity that can be computed from observable variables of the system to provide or aid prognosis. Features can be also referred to as a 1xN feature vector $F^l(i)$ of the lth UUT at time index i.

Operational Conditions: Let $c_m^l(i)$ be an operational condition at time index i, where $m = 1, 2, ..., M$ is the condition index, and $l = 1, 2, ..., L$ is the UUT index.

Operational conditions describe how the system is being operated and also include the load on the system. The conditions can also be referred to as a 1xM vector $C^l(i)$ of the lth UUT at time index i. The matrix C^l for all times $< t_P$ is referred to as load history and for times $\geq t_P$ as operational (load) profile for the system.

Health Index: Let $h^l(i)$ be a health index at time index i for UUT $l = 1, 2, ..., L$. h can be considered a normalized aggregate of health indicators (relevant features) and operational conditions.

Ground Truth: Ground truth, denoted by the subscript $*$, represents the best belief about the true value of a system variable. In the feature domain $f_{*n}^l(i)$ may be directly or indirectly calculated from measurements. In the health domain, $h_*^l(i)$ is the computed health at time index i for UUT $l = 1, 2, ..., L$ after a run-to-failure test. For an offline study EoL_* is the known end-of-life point for the system.

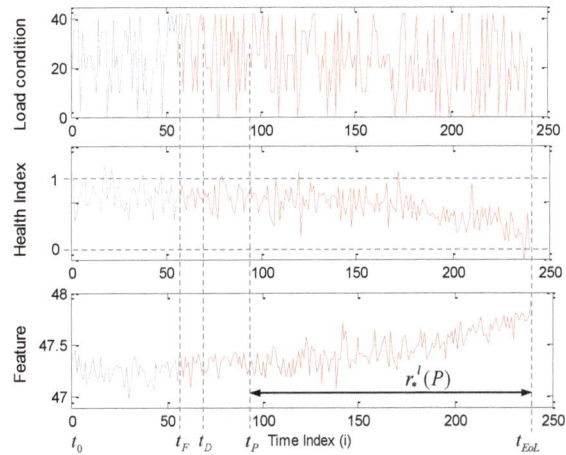

Figure 5: Features and conditions for l^{th} UUT (Saxena, et al., 2008).

History Data: History data, denoted by the subscript $\#$, encapsulates all the a priori information we have about a system. Such information may be of the form of archived measurements or observed EoL data, and can refer to variables in both the feature and health domains represented by $f_{\#n}^l(i)$ and $h_\#^l(i)$ respectively. For a fleet of systems all reliability estimates such as MTTF or MTBF would be considered history data.

Point Prediction: Let $\phi^l(i \mid j)$ be a prediction for a variable of interest at a desired point of time t_j given information up to time t_j, where $t_j \leq t_i$ (see Figure 6). Predictions can be made in any domain, features or health. In some cases it is useful to extrapolate features and then aggregate them to compute health and in other cases features are aggregated to a health and then extrapolated to estimate RUL. It must be noted here that a point prediction may be expressed as probability a distribution or estimated moments derived from the probability distribution.

Trajectory Prediction: Let $\underline{\Phi}^l(i)$ be a trajectory of predictions formed by point predictions for a variable of interest from time index i onwards such that $\underline{\Phi}^l(i) = \{\phi^l(i|i), \phi^l(i+1|i), ..., \phi^l(EoP|i)\}$ (see Figure 6). It must be noted that only the last point of this trajectory, i.e., $\phi^l(EoP|i)$ is used to estimate RUL.

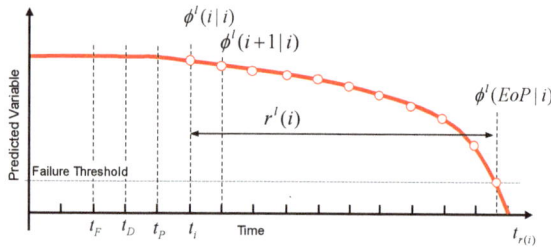

Figure 6: Illustration showing a trajectory prediction. Predictions get updated every time instant.

RUL: Let $r^l(i)$ be the remaining useful life estimate at time index i given that the information (features and conditions) up to time index i and an expected operational profile for the future are available. RUL is computed as the difference between the predicted time of failure (where health index approaches zero) and the current time t_i. RUL is estimated as

$$r^l(i) = t_j - t_i, \text{ where } j = \max_z \{h^l(z) \geq 0, z > i\}. \quad (1)$$

Corresponding ground truth is computed as

$$r_*^l(i) = t_j - t_i, \text{ where } j = \max_z \{h_*^l(z) \geq 0, z > i\}. \quad (2)$$

RUL vs. Time Plot: RUL values are plotted against time to compare with RUL ground truth (represented by a straight line). As illustrated in Figure 7, this visually summarizes prediction performance as it evolves through time. This plot is the foundation of prognostic metrics developed in subsequent sections.

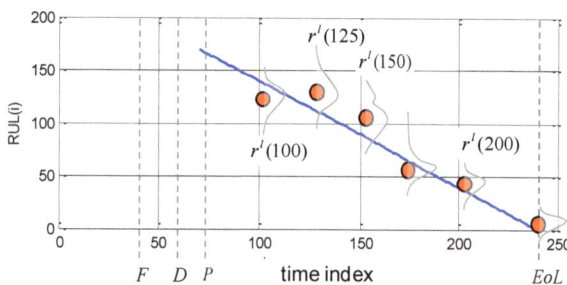

Figure 7: Comparing RUL predictions from ground truth ($p = \{P \mid P \in [70,240]\}$, $t_{EoL} = 240$, $t_{EoP} > 240$) (Saxena, et al., 2008).

5.1 Incorporating Uncertainty Estimates

As discussed in section 4, prognostics is meaningless unless the uncertainties in the predictions are accounted for. PAs can handle these uncertainties in various ways such as propagating through time the prior probabilities of uncertain inputs and estimating posteriori distributions of EoL and RUL quantities (Orchard & Vachtsevanos, 2009). Therefore, the metrics should be designed such that they can make use of these distributions while assessing the performance. The first step in doing so is to define a reasonable point estimate from these distributions such that no interesting features get ignored in decision making. Computationally the simplest, and hence most widely used, practice has been to compute mean and variance estimates of these distributions (Goebel, Saha, & Saxena, 2008). In reality these distributions are rarely smooth or symmetric, thereby resulting in large errors due to such simplifying assumptions especially while carrying out performance assessment. It is, therefore, suggested that other estimates of central tendency (location) and variance (spread) be used instead of mean and standard deviation, which are appropriate only for Normal cases. For situations were normality of the distribution cannot be established, it is preferable to rely on median as a measure of location and the quartiles or Inter Quartile Range (IQR) as a measure of spread (Hoaglin, Mosteller, & Tukey, 1983). Various types of distributions are categorized into four categories and corresponding methods to compute more appropriate location and spread measures are suggested in Table 3. For the purpose of plotting and visualizing the data use of error bars and box-plots is suggested (Figure 8); more explanation is given in the following sections.

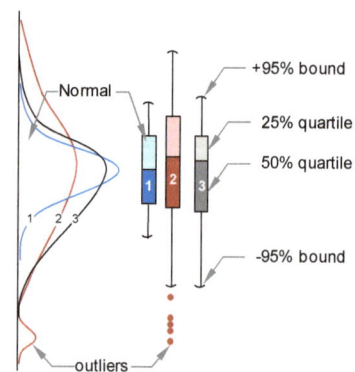

Figure 8: Visual representation for distributions. Distributions shown on the left can be represented by box plots as shown on the right (Saxena, et al., 2009b).

While mean and variance estimates are good for easy understanding they can be less robust when deviations from assumed distribution category are random and frequent. Furthermore, given the fact that there will be uncertainty in any prediction one must make provisions to account for these deviations. One common way to do so is to specify an allowable error bound around the point of interest and one could use the total probability of failure within that error bound instead of basing a decision on a single point estimate. As shown in Figure 9, this error bound may be asymmetric especially in the case of prognostics, since it is often argued that an early prediction is preferred over a late prediction.

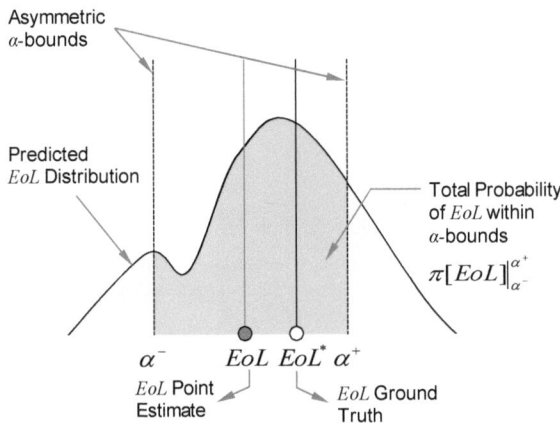

Figure 9: Concepts for incorporating uncertainties.

These ideas can be analytically incorporated into the numerical aspect of the metrics by computing the probability mass of a prediction falling within the specified α-bounds. As illustrated in the figure, the EoL ground truth may be very different than the estimated EoL and hence the decisions based on probability mass are expectedly more robust. Computing the probability mass requires integrating the probability distribution between the α-bounds (Figure 10).

The cases where analytical form of the distribution is available, like for Normal distributions, this probability mass can be computed analytically by integrating the area under the prediction PDF between the α-bounds (α^- to α^+). However, for cases where there is no analytical form available, a summation based on histogram obtained from the process/algorithm can be used to compute this probability (see Figure 10). A formal way to include this probability mass into the analytical framework is by introducing a β-criterion, where a prediction is considered inside α-bounds only if the probability mass of the corresponding distribution within the α-bounds is more than a predetermined threshold β. This parameter is also linked to the issues of uncertainty management and risk absorbing capacity of the system.

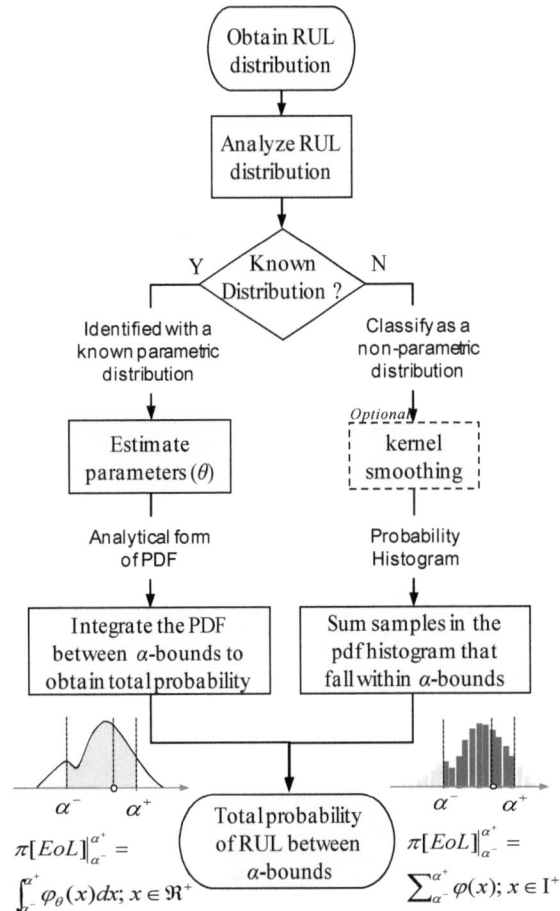

Figure 10: Procedure to compute probability mass of RULs falling within specified α-bounds.

The categorization shown in Table 3 determines the method of computing the probability of RULs falling between α-bounds, i.e., area integration or discrete summation, as well as how to represent it visually. For cases that involve a Normal distribution, using a confidence interval represented by a confidence bar around the point prediction is sufficient (Devore, 2004). For situations with non-Normal single mode distributions this can be done with an inter-quartile plot represented by a box plot (Martinez, 2004). Box plots convey how a prediction distribution is skewed and whether this skew should be considered while computing a metric. A box plot also has provisions to represent outliers, which may be useful to keep track of in risk sensitive situations. It is suggested to use box plots superimposed with a dot representing the mean of the distribution. This will allow keeping the visual information in perspective with respect to the conventional plots. For the mixture of Gaussians case, it is recommended that a model with few (preferably $n \leq 4$) Gaussian modes is created and corresponding confidence bars plotted adjacent to each other. The weights for each Gaussian component can then be represented by the thickness of the error bars. It is not recommended to plot multiple box plots since there is no methodical way to differentiate and isolate the samples associated to individual Gaussian components, and compute the quartile ranges separately for each of them. A linear additive model is assumed here for simplicity while computing the mixture of Gaussians.

$$\varphi(x) \cong \omega_1 \cdot N(\mu_1, \sigma_1) + \ldots + \omega_n \cdot N(\mu_n, \sigma_n); n \in I^+ \quad (3)$$

where:

$\varphi(x)$ is a PDF with of multiple Gaussians

ω is the weight factor for each Gaussian component

$N(\mu, \sigma)$ is a Gaussian distribution with parameters μ and σ

n is the number of Gaussian modes identified in the distribution.

Table 3: Methodology to select location and spread measures along with visualization methods (Saxena, et al., 2009b).

	Normal Distribution	Mixture of Gaussians	Non-Normal Distribution	Multimodal (non-Normal)
	Parametric		Non-Parametric	
Location (Central tendency)	Mean (μ)	Means: $\mu_1, \mu_2, \ldots, \mu_n$ weights: $\omega_1, \omega_2, \ldots, \omega_n$	Mean, Median, L-estimator, M-estimator	Dominant median, Multiple medians, L-estimator, M-estimator
Spread (variability)	Sample standard deviation (σ), IQR (inter quartile range)	Sample standard deviations: $\sigma_1, \sigma_2, \ldots, \sigma_n$	Mean Absolute Deviation (MAD), Median Absolute Deviation (MdAD), Bootstrap methods, IQR	
Visualization	Confidence Interval (CI), Box plot with mean	Multiple CIs with varying bar width Note: here $\omega_1 > \omega_2 > \omega_3$	Box plot with mean	Box plot with mean

6. PERFORMANCE METRICS

6.1 Limitations of Classical Metrics

In Saxena, et al. (2009a) it was reported that the most commonly used metrics in the forecasting applications are accuracy (bias), precision (spread), MSE, and MAPE. Tracking the evolution of these metrics one can see that these metrics were successively developed to incorporate issues not covered by their predecessors. There are more variations and modifications that can be found in literature that measure different aspects of performance. Although these metrics captured important aspects, this paper focuses on enumerating various shortcomings of these metrics from a prognostics viewpoint. Researchers in the PHM community have further adapted these metrics to tackle

these shortcomings in many ways (Saxena, et al., 2008). However, there are some fundamental differences between the performance requirements from general forecasting applications and prognostics applications that did not get adequately addressed. This translates into differences at the design level for the metrics in either case. Some of these differences are discussed here.

These metrics provide a statistical accounting of variations in the distribution of RULs. Whereas this is meaningful information, these metrics are not designed for applications where RULs are continuously updated as more data becomes available. Prognostics prediction performance (e.g., accuracy and precision) tends to be more critical as time passes by and the system nears its end-of-life. Considering EoL as a fixed reference point in time, predictions made at different times create several conceptual difficulties in computing an aggregate measure using conventional metrics. Predictions made early on have access to less information about the dynamics of fault evolution and are required to predict farther in time. This makes the prediction task more difficult as compared to predicting at a later stage. Each successive prediction utilizes additional data available to it. Therefore, a simple aggregate of performance over multiple predictions made is not a fair representative of overall performance. It may be reasonable to aggregate fixed n-step ahead (fixed horizon) predictions instead of aggregating EoL predictions (moving horizon). Performance at specific times relative to the EoL can be a reasonable alternative as well. Furthermore, most physical processes describing fault evolution tend to be more or less monotonic in nature. In such cases it becomes easier to learn true parameters of the process as more data become available. Thus, it may be equally important to quantify how well and how quickly an algorithm improves as more data become available.

Following from the previous argument, conventional measures of accuracy and precision tend to account for statistical bias and spread arising from the system. What is missing from the prognostics point of view is a measure that encapsulates the notion of performance improvement with time, since prognostics continuously updates, i.e., successive predictions occur at early stages close to fault detection, middle stages while the fault evolves, and late stages nearing EoL. Depending on application scenarios, criticality of predictions at different stages may be ranked differently. A robust metric should be capable of making an assessment at all stages. This will not only allow ranking various algorithms at different stages but also allow switching prediction models with evolving fault stages instead of using a single prediction algorithm until EoL.

Time scales involved in prognostics applications vary widely (on the order of seconds and minutes for electronic components vs. weeks and years for battery packs). This raises an important question - "how far in advance is enough when predicting with a desired confidence?" Although the earlier the better, a sufficient time to plan and carry out an appropriate corrective action is what is sought. While qualitatively these performance measures remain the same (i.e., accuracy and precision) one needs to incorporate the issues of time criticality.

The new metrics developed and discussed in the following sections attempt to alleviate some of these issues in evaluating prognostic performance.

6.2 Prognostic Performance Metrics

In this paper four metrics are discussed that can be used to evaluate prognostic performance while keeping in mind the various issued discussed earlier. These four metrics follow a systematic progression in terms of the information they seek (Figure 11).

The first metric, Prognostic Horizon, identifies whether an algorithm predicts within a specified error margin (specified by the parameter α, as discussed in the section 5.1) around the actual EoL and if it does how much time it allows for any corrective action to be taken. In other words it assesses whether an algorithm yields a sufficient prognostic horizon; if not, it may not be meaningful to continue on computing other metrics. If an algorithm passes the PH test, the next metric, α-λ Performance, goes further to identify whether the algorithm performs within desired error margins (specified by the parameter α) of the actual RUL at any given time instant (specified by the parameter λ) that may be of interest to a particular application. This presents a more stringent requirement of staying within a converging cone of the error margin as a system nears EoL. If this criterion is also met, the next step is to quantify the accuracy levels relative to the actual RUL. This is accomplished by the metrics Relative Accuracy and Cumulative Relative Accuracy. These metrics assume that prognostic performance improves as more information becomes available with time and hence, by design, an algorithm will satisfy these metrics criteria if it converges to true RULs. Therefore, the fourth metric, Convergence, quantifies how fast the algorithm converges if it does satisfy all previous metrics. These metrics can be considered as a hierarchical test that provides several levels of comparison among different algorithms in addition to the specific information these metrics individually provide regarding algorithm performance.

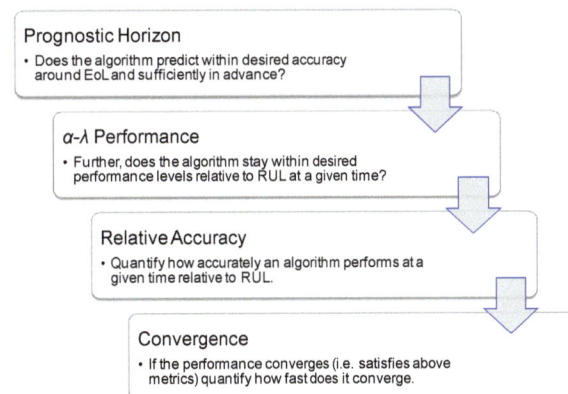

Figure 11: Hierarchical design of the prognostics metrics.

It must be noted that these metrics share the attribute of performance tracking with time unlike the classical metrics. Discussion on detailed definitions and descriptions of these metrics follows henceforth.

Prognostic Horizon: Prognostic Horizon (PH) is defined as the difference between the time index i when the predictions first meet the specified performance criteria (based on data accumulated until time index i) and the time index for EoL. The performance requirement may be specified in terms of an allowable error bound (α) around the true EoL. The choice of α depends on the estimate of time required to take a corrective action. Depending on the situation this corrective action may correspond to performing maintenance (manufacturing plants) or bringing the system to a safe operating mode (operations in a combat zone).

$$PH = t_{EoL} - t_{i_{\alpha\beta}} \qquad (4)$$

where:

$i_{\alpha\beta} = \min\left\{ j \mid (j \in p) \wedge \left(\pi[r(j)]_{-\alpha}^{+\alpha} \geq \beta \right) \right\}$ is the first time index when predictions satisfy β-criterion for a given α

p is the set of all time indexes when predictions are made

l is the index for l^{th} unit under test (UUT)

β is the minimum acceptable probability mass

$r(j)$ is the predicted RUL distribution at time t_j

t_{EoL} is the predicted End-of-Life

$\pi[r(j)]_{\alpha^-}^{\alpha^+}$ is the probability mass of the prediction PDF within the α-bounds that are given by $\alpha^+ = r_* + \alpha \cdot t_{EoL}$ and $\alpha^- = r_* - \alpha \cdot t_{EoL}$

As shown in Figure 12, the desired level of accuracy with respect to the EoL ground truth is specified as $\pm\alpha$-bounds (shaded band). RUL distributions are then plotted against time for all the algorithms that are to be compared. In simple cases the evaluation may be based on point estimates (mean, median, etc.) of the distributions. The PH for an algorithm is declared as soon the corresponding prediction enters the band of desired accuracy. As is evident from the illustration in Figure 12(a), the second algorithm (A2) has a longer PH. However, looking closely at the plots, A1 does not perform much worse than A2, but this method, being less robust due to use of only a point estimate, results in very different PH values for the two algorithms. This can be improved by using the β-criterion, as shown in Figure 12(b).

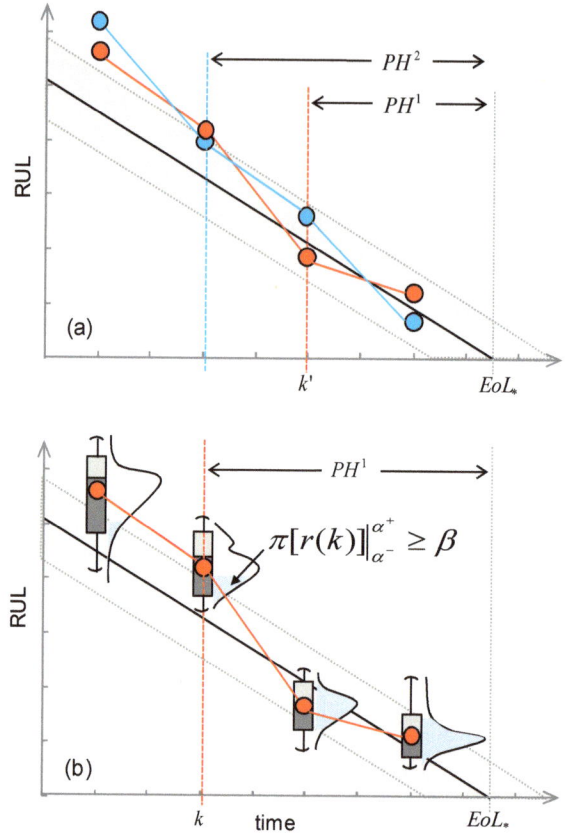

Figure 12: (a) Illustration of Prognostics Horizon while comparing two algorithms based on point estimates (distribution means) (b) PH based on β-criterion results in a more robust metric.

Prognostic horizon produces a score that depends on length of ailing life of a system and the time scales in the problem at hand. The range of PH is between (t_{EoL}-t_P) and $max\{0, t_{EoL}-t_{EoP}\}$. The best score for PH is obtained when an algorithm always predicts within desired accuracy zone and the worst score when it never predicts within the accuracy zone. The notion for Prediction Horizon has been long discussed in the literature from a conceptual point of view. This metric indicates whether the predicted estimates are within specified limits around the actual EoL so that the predictions are considered trustworthy. It is clear that a longer prognostic horizon results in more time available to act based on a prediction that has some desired credibility. Therefore, when comparing algorithms, an algorithm with longer prediction horizon would be preferred.

α-λ Performance: This metric quantifies prediction quality by determining whether the prediction falls within specified limits at particular times with respect to a performance measure. These time instances may be specified as percentage of total ailing life of the system. The discussion henceforth is presented in the context of accuracy as a performance measure, hence α-λ *accuracy,* but any performance measure of interest may fit in this framework.

α-λ *accuracy* is defined as a binary metric that evaluates whether the prediction accuracy at specific time instance t_λ falls within specified α-bounds (Figure 13). Here t_λ is a fraction of time between t_P and the actual t_{EoL}. The α-bounds here are expressed as a percentage of actual RUL $r(i_\lambda)$ at t_λ.

$$\alpha - \lambda \ Accuracy = \begin{cases} 1 & \text{if} \quad \pi[r(i_\lambda)]_{-\alpha}^{+\alpha} \geq \beta \\ 0 & \text{otherwise} \end{cases} \quad (5)$$

where:

λ is the time window modifier such that $t_\lambda = t_P + \lambda(t_{EoL} - t_P)$

β is the minimum acceptable probability for β-criterion

$r(i_\lambda)$ is the predicted RUL at time index i_λ

$\pi[r(i_\lambda)]_{\alpha^-}^{\alpha^+}$ is the probability mass of the prediction PDF within the α-bounds that are given by $\alpha^+ = r_*(i_\lambda) + \alpha \cdot r(i_\lambda)$ and $\alpha^- = r_*(i_\lambda) - \alpha \cdot r(i_\lambda)$

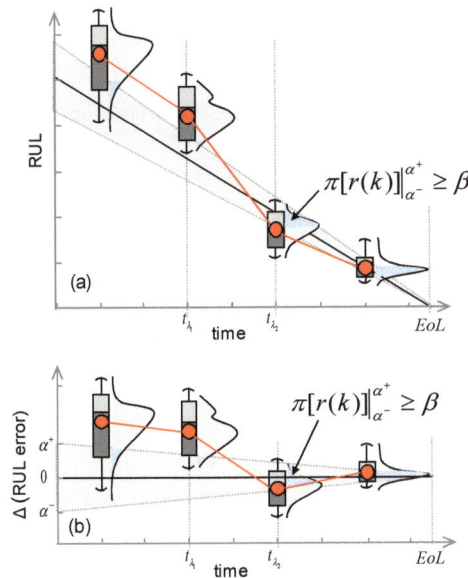

Figure 13: (a) α-λ accuracy with the accuracy cone shrinking with time on RUL vs. time plot. (b) Alternate representation of α-λ accuracy on RUL-error vs. time plot.

As an example, this metric would determine whether a prediction falls within 10% accuracy (α = 0.1) of the true RUL halfway to failure from the time the first prediction is made (λ = 0.5). The output of this metric is binary (1=Yes or 0=No) stating whether the

desired condition is met at a particular time. This is a more stringent requirement as compared to prediction horizon, as it requires predictions to stay within a cone of accuracy i.e., the bounds that shrink as time passes by as shown in Figure 13(a). For easier interpretability α-λ accuracy can also be plotted as shown in Figure 13(b). It must be noted that the set of all time indexes (p) where a prediction is made is determined by the frequency of prediction step in a PA. Therefore, it is possible that for a given λ there is no prediction assessed at time t_λ if the corresponding $i_\lambda \notin p$. In such cases one can make alternative arrangements such as choosing another λ' closest to λ such that $i_{\lambda'} \in p$.

Relative Accuracy: Relative Accuracy (RA) is defined as a measure of error in RUL prediction relative to the actual RUL $r_*(i_\lambda)$ at a specific time index i_λ.

$$RA_\lambda^l = 1 - \frac{\left| r_*^l(i_\lambda) - \langle r^l(i_\lambda) \rangle \right|}{r_*^l(i_\lambda)} \quad (6)$$

where:

λ is the time window modifier such that $t_\lambda = t_P + \lambda(t_{EoL} - t_P)$,

l is the index for l^{th} unit under test (UUT),

$r_*(i_\lambda)$ is the ground truth RUL at time index i_λ,

$\langle r(i_\lambda) \rangle$ is an appropriate central tendency point estimate of the predicted RUL distribution at time index i_λ.

This is a notion similar to α-λ *accuracy* where, instead of finding out whether the predictions fall within a given accuracy level at a given time instant, accuracy is measured quantitatively (see Figure 14). First a suitable central tendency point estimate is obtained from the prediction probability distribution using guidelines provided in Table 3 and then using Eq.6.

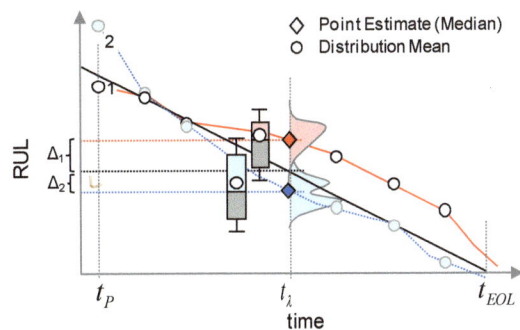

Figure 14: Schematic illustrating Relative Accuracy.

RA may be computed at a desired time t_λ. For cases with mixture of Gaussians a weighted aggregate of the means of individual modes can be used as the point estimate; where the weighting function is the same as the one for the various Gaussian components in the distribution. An algorithm with higher relative accuracy is desirable. The range of values for RA is [0,1], where the perfect score is 1. It must be noted that if the prediction error magnitude grows beyond 100%, RA results in a negative value. Large errors like these, if interpreted in terms of α parameter for previous metrics, would correspond to values greater than 1. Cases like these need not be considered as it is expected that, under reasonable assumptions, preferred α values will be less than 1 for PH and α-λ accuracy metrics and that these cases would not have met those criteria anyway.

RA conveys information at a specific time. It can be evaluated at multiple time instances before t_λ to account for general behavior of the algorithm over time. To aggregate these accuracy levels, Cumulative Relative Accuracy (CRA) can be defined as a normalized weighted sum of relative accuracies at specific time instances.

$$CRA_\lambda^l = \frac{1}{|p_\lambda|} \sum_{i \in p_\lambda} w(r^l(i)) RA_\lambda^l \qquad (7)$$

where:

$w(r^l(i))$ is a weight factor as a function of RUL at all time indices

p_λ is the set of all time indexes before t_λ when a prediction is made

$|p_\lambda|$ is the cardinality of the set

In most cases it is desirable to weigh those relative accuracies higher that are closer to t_{EoL}. In general, it is expected that t_λ is chosen such that it holds some physical significance such as a time index that provides a required prediction horizon, or time required to apply a corrective action, etc. For instance, RA evaluated at $t_{0.5}$ signifies the time when a system is expected to have consumed half of its ailing life, or in terms of damage index the time index when damage magnitude has reached 50% of the failure threshold. This metric is useful in comparing different algorithms for a given λ in order to get an idea on how well a particular algorithm does at significant times. Choice of t_λ should also take into account the uncertainty levels that an algorithm entails by making sure that the distribution spread at t_λ does not cross over expected t_{EoL} by significant margins especially for critical applications. In other words the probability mass of the RUL

distribution at t_λ extending beyond EoL should not be too large.

Convergence: Convergence is a meta-metric defined to quantify the rate at which any metric (M) like accuracy or precision improves with time. It is defined as the distance between the origin and the centroid of the area under the curve for a metric is a measure of convergence rate.

$$C_M = \sqrt{(x_c - t_P)^2 + y_c^2}, \qquad (8)$$

where:

C_M is the Euclidean distance between the center of mass (x_c, y_c) and $(t_P, 0)$

$M(i)$ is a non-negative prediction accuracy or precision metric with a time varying value

(x_c, y_c) is the center of mass of the area under the curve $M(i)$ between t_P and t_{EoUP}, defined as following

$$x_c = \frac{\frac{1}{2} \sum_{i=P}^{EoUP} (t_{i+1}^2 - t_i^2) M(i)}{\sum_{i=P}^{EoUP} (t_{i+1} - t_i) M(i)}, \text{ and}$$

$$y_c = \frac{\frac{1}{2} \sum_{i=P}^{EoUP} (t_{i+1} - t_i) M(i)^2}{\sum_{i=P}^{EoUP} (t_{i+1} - t_i) M(i)}. \qquad (9)$$

As suggested earlier, this discussion assumes that the algorithm performance improves with time. This is easily established if it has passed criteria for previous metrics. For illustration of the concept in Figure 15 three cases are shown that converge at different rates. Lower distance means a faster convergence. Convergence is a useful metric since we expect a prognostics algorithm to converge to the true value as more information accumulates over time. Further, a faster convergence is desired to achieve a high confidence in keeping the prediction horizon as large as possible.

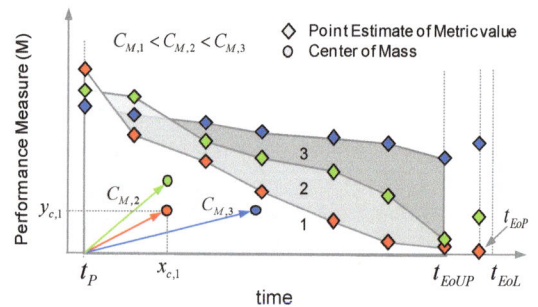

Figure 15: Convergence compares the rates at which different algorithms improve.

6.2.1 Applying the Prognostics Metrics

In practice, there can be several situations where the definitions discussed above result in ambiguity. In Saxena, et al. (2009a) several such situations have been discussed in detail with corresponding suggested resolutions. For the sake of completeness such situations are very briefly discussed here.

With regards to PH metric, the most common situation encountered is when the RUL trajectory jumps out of the $\pm\alpha$ accuracy bounds temporarily. Situations like this result in multiple time indexes where RUL trajectory enters the accuracy zone to satisfy the metric criteria. A simple and conservative approach to deal with this situation is to declare a PH at the latest time instant the predictions enter accuracy zone. Another option is to use the original PH definition and further evaluate other metrics to determine whether the algorithm satisfies all other requirements. Situations like these can occur due to a variety of reasons.

- *Inadequate system model:* Real systems often exhibit inherent transients at different stages during their life cycles. These transients get reflected as deviations in computed RUL estimates from the true value if the underlying model assumed for the system does not account for these behaviors. In such cases, one must step back and refine the respective models to incorporate such dynamics.

- *Operational transients:* Another source of such behaviors can be due to sudden changes in operational profiles under which a system is operating. Prognostic algorithms may show a time lag in adapting to such changes and hence resulting in temporary deviation from the real values.

- *Uncertainties in prognostic environments:* Prognostics models a stochastic process and hence the behavior observed from a particular run (single realization of the stochastic process) may not exhibit the true nature of prediction trajectories. Assuming that all possible measures for uncertainty reduction have been taken during algorithm development, such observations should be treated as isolated realization of the process. In that case these trajectories should be aggregated from multiple runs to achieve statistical significance or more sophisticated stochastic analyses can be carried out.

Plotting the RUL trajectory in the PH plot provides insights for such deficiencies to algorithm developers. It is important to identify the correct reason before computing a metric and interpreting its result. Ideally, an algorithm and a system model should be robust to transients inherent to the system behavior and operational conditions.

The situations discussed above are more common towards the end when a system nears EoL. This is because in most cases the fault evolution dynamics are too fast and complex to model or learn from data as the system nears EoL. Therefore, RUL curve deviates from the error band near t_{EoL}. To determine whether such deviations are critical for post-prognostic decision making, the concept of t_{EoUP} or End-of-Useful-Predictions (EoUP) is introduced. This index represents the minimum allowable PH that is required to take a corrective measure. Any predictions made beyond EoUP are of little or no use from a practical viewpoint.

6.2.2 Choosing Performance Parameters

From a top-down perspective, the main idea behind these metrics is to help management generate appropriate specifications and requirements for prognostics algorithm in fielded applications. The outcome of these metrics depends directly on the values chosen for input parameters like α, λ, and β. Thus, the choice of values for these parameters forms an important aspect of performance evaluation and interpretation. Cost-benefit-risk analyses are generally performed through various methods that model the effects of a variety of constraints (financial costs, safety, criticality of mission completion, reputation, etc.) and derive a range of available slacks in achieving an overall benefit situation (pareto optimal solutions). It is expected that the parameters can be incorporated in these analyses to include the effects of prognostic performance on the cost-benefit of PHM. While this subject is out of the scope of this paper a brief discussion is provided for an overall perspective on how these parameters can be connected to safety, logistics and cost constraints.

There are systems that involve different levels of criticality when they fail. In a mission critical scenario a failure may be catastrophic and hence a limited number of false positives may be tolerable but no false negatives. In other cases the cost of acting on false positives may be prohibitively high. There are cases where it is more cost effective to tolerate several false negatives as opposed to reacting to a false positive and hence it is acceptable even if the system runs to failure once in a while. There are several factors that determine how critical it may be to make a correct prediction. These factors combined together should dictate the choice of these parameters while carrying out performance evaluation. Some of the most important factors are:

- *Time for problem mitigation*: the amount of time to mitigate a problem or start a corrective action when critical health deterioration of a component/system has been detected is a very important factor. As mentioned earlier, very accurate predictions at a time when no recovery action can be made are not useful. Hence, a tradeoff between error tolerance and time for recovery from fault should be considered. The time for problem mitigation will vary from system to system and involves multiple factors. This factor will have a direct consequence on λ parameter.

- *Cost of mitigation*: cost of the reparative action is an important factor in all management related decisions and hence should be considered. From a decision making point of view this can be associated to the cost due to false positives. This factor influences α, where there is often a tradeoff between false positives and true positive rates.

- *Criticality of system or cost of failure*: This quantifies the effect of false negatives. Further, while comparing time-critical scenarios, resources should be directed towards more critical and important components in order to efficiently maintain overall health of the system. Likewise, if the health assessment is being performed on multiple units in a system, the parameters for different units should be chosen based on a prioritized list of criticality. Assessment of criticality is usually done based on severity and frequency of occurrence statistics available from Failure Modes, Effects, and Criticality Analysis (FMECA) studies (MIL-STD-1629A, 1980). Another perspective to assess criticality is based on cost-benefit analysis where cost of failures is incorporated to assess the implications of false negatives (Banks & Merenich, 2007; Feldman, et al., 2008).

- *Uncertainty management capability*: Level of confidence on the uncertainty management capability and costs of system failure determine the risk absorbing capacity in a particular scenario. The choice of β is guided by such factors.

Note that these factors mentioned here are not arranged based on any order of importance; users should consider them based on the characteristics of their systems and may skip a few as appropriate.

7. FUTURE WORK

A natural extension of this work leads into the development of online prognostic performance metrics. This would require investigations into several issues that were set aside through various assumptions in the present work. For instance, thus far performance evaluation ignores the effect of future loading conditions that alter the rate of remaining life consumption. Performance evaluation without an explicit knowledge about EoL is a challenge for online metrics. These metrics will also need to include provisions for the effects of scheduled maintenance and

self-healing characteristics in some systems. Further, the concepts presented in this paper will be refined and applied to a variety of applications. Developing more metrics like robustness and sensitivity, etc. also remains on the research agenda. Finally, a formal framework for connecting these metrics to top level requirements through development of uncertainty management and representation (URM) methods, incorporation of risk analysis, cost-benefit analysis, and requirements flow down remains a topic of interest in future work.

8. CONCLUSION

This paper presents several performance metrics for offline evaluation of prognostics algorithms. A brief overview of different methods employed for performance evaluation is also included. It has been shown that various forecasting related applications differ from prognostics in the systems health management context. This called for developing specialized metrics for prognostics. These metrics were developed keeping in mind various critical aspects that must be included in performance evaluation. A formal prognostic framework was presented to clearly define the concepts and introduce the terminology. Metrics with uncertainty representation capabilities were developed that track the performance of an algorithm with time. Along with detailed discussions and illustrations, it has been shown that these metrics can be successfully applied to evaluate prognostic performance in a standardized manner. Furthermore, it has been discussed that the suggested metrics can be employed to reflect high level requirements in a practical PHM system.

ACKNOWLEDGMENT

The authors would like to express their gratitude to colleagues at the Prognostic Center of Excellence (NASA Ames Research Center) and external partners at Impact Technologies and Clarkson University for participating in research discussions, evaluating metrics in their respective applications, and providing a valuable feedback. This work was funded by NASA Aviation Safety Program-IVHM Project.

REFERENCES

Banks, J., & Merenich, J. (2007). *Cost benefit analysis for asset health management technology*. Reliability and Maintainability Symposium (RAMS), Orlando, FL.

Carrasco, M., & Cassady, C. R. (2006). *A study of the impact of prognostic errors on system performance*. Annual Reliability and Maintainability Symposium, RAMS06.

Coble, J. B., & Hines, J. W. (2008). *Prognostic Algorithm Categorization with PHM Challenge*

Application. 1st International Conference on Prognostics and Health Management (PHM08), Denver, CO.

Coppe, A., Haftka, R. T., Kim, N., & Yuan, F. (2009). *Reducing Uncertainty in Damage Growth Properties by Structural Health Monitoring.* Annual Conference of the Prognostics and Health Management Society (PHM09) San Diego, CA.

DeNeufville, R. (2004). *Uncertainty Management for Engineering Systems Planning and Design.* Engineering Systems Symposium MIT, Cambridge, MA.

Devore, J. L. (2004). *Probability and Statistics for Engineering and the Sciences* (6th ed.): Thomson.

Drummond, C., & Yang, C. (2008). *Reverse Engineering Costs: How Much will a Prognostic Algorithm Save?* International Conference on Prognostics and Health Management, Denver, CO.

Engel, S. J. (2008). *Prognosis Requirements and V&V: Panel Discussion on PHM Capabilities: Verification, Validation, and Certification Issues.* International Conference on Prognostics and Health Management (PHM08), Denver, CO.

Engel, S. J., Gilmartin, B. J., Bongort, K., & Hess, A. (2000). *Prognostics, the Real Issues Involved with Predicting Life Remaining.* IEEE Aerospace Conference, Big Sky, MT.

Feldman, K., Sandborn, P., & Jazouli, T. (2008). *The Analysis of Return on Investment for PHM Applied to Electronic Systems.* International Conference on Prognostics and Health Management (PHM08), Denver, CO.

Goebel, K., & Bonissone, P. (2005). *Prognostic Information Fusion for Constant Load Systems.* 7th Annual Conference on Information Fusion.

Goebel, K., Saha, B., & Saxena, A. (2008). *A Comparison of Three Data-Driven Techniques for Prognostics.* 62nd Meeting of the Society For Machinery Failure Prevention Technology (MFPT), Virginia Beach, VA.

Guan, X., Liu, Y., Saxena, A., Celaya, J., & Goebel, K. (2009). *Entropy-Based Probabilistic Fatigue Damage Prognosis and Algorithmic Performance Comparison.* Annual Conference of the Prognostics and Health Management Society (PHM09), San Diego, CA.

Hastings, D., & McManus, H. (2004). *A Framework for Understanding Uncertainty and its Mitigation and Exploitation in Complex Systems.* Engineering Systems Symposium MIT, Cambridge MA.

Hoaglin, D. C., Mosteller, F., & Tukey, J. W. (Eds.). (1983). *Understanding Robust and Exploratory Data Analysis*: John Wiley & Sons.

ISO (2004). Condition Monitoring and Diagnostics of Machines - Prognostics part 1: General Guidelines, ISO/IEC Directives Part 2 C.F.R..

Leao, B. P., Yoneyama, T., Rocha, G. C., & Fitzgibbon, K. T. (2008). *Prognostics Performance Metrics and Their Relation to Requirements, Design, Verification and Cost-Benefit.* International Conference on Prognostics and Health Management (PHM08), Denver CO.

Martinez, A. R. (2004). Exploratory Data Analysis with MATLAB. In A. R. Martinez (Ed.): CRC Press.

MIL-STD-1629A. (1980). Military Standard: Procedures for Performing A Failure Mode, Effects and Criticality Analysis. Washington DC: Department of Defense.

NASA. (2009). NASA Aviation Safety Program Retrieved December 2009, from http://www.aeronautics.nasa.gov/programs_avsafe.htm

Ng, K.-C., & Abramson, B. (1990). Uncertainty Management in Expert Systems. *IEEE Expert Systems, 5,* 20.

Orchard, M., Kacprzynski, G., Goebel, K., Saha, B., & Vachtsevanos, G. (2008). *Advances in Uncertainty Representation and Management for Particle Filtering Applied to Prognostics.* International Conference on Prognostics and Health Management (PHM08), Denver, CO.

Orchard, M. E., Tang, L., Goebel, K., & Vachtsevanos, G. (2009). *A Novel RSPF Approach to Prediction of High-Risk, Low-Probability Failure Events.* Annual Conference of the Prognostics and Health Management Society (PHM09), San Diego, CA.

Orchard, M. E., & Vachtsevanos, G. J. (2009). A Particle-Filtering Approach for On-line Fault Diagnosis and Failure Prognosis. *Transactions of the Institute of Measurement and Control, 31*(3-4), 221-246.

Orsagh, R. F., Roemer, M. J., Savage, C. J., & McClintic, K. (2001). *Development of Effectiveness and Performance Metrics for Mechanical Diagnostic Techniques.* 55th Meeting of the Society for Machinery Failure Prevention Technology, Virginia Beach, VA.

Pipe, K. (2008). *Practical Prognostics for Condition Based Maintenance.* International Conference on Prognostics and Health Management (PHM08), Denver, CO.

Saha, B., & Goebel, K. (2009). *Modeling Li-ion Battery Capacity Depletion in a Particle Filtering Framework.* Annual Conference of the Prognostics and Health Management Society (PHM09), San Diego, CA.

Sankararaman, S., Ling, Y., Shantz, C., & Mahadevan, S. (2009). *Uncertainty Quantification in Fatigue Damage Prognosis.* Annual Conference of the Prognostics and Health Management Society (PHM09), San Diego, CA.

Saxena, A., Celaya, J., Balaban, E., Goebel, K., Saha, B., Saha, S., et al. (2008). *Metrics for Evaluating Performance of Prognostics Techniques.* 1st International Conference on Prognostics and Health Management (PHM08), Denver, CO.

Saxena, A., Celaya, J., Saha, B., Saha, S., & Goebel, K. (2009a). *Evaluating Algorithmic Performance Metrics Tailored for Prognostics.* IEEE Aerospace Conference, Big Sky, MT.

Saxena, A., Celaya, J., Saha, B., Saha, S., & Goebel, K. (2009b). *On Applying the Prognostics Performance Metrics.* Annual Conference of the Prognostics and Health Management Society (PHM09) San Diego, CA.

Schwabacher, M. (2005). *A Survey of Data Driven Prognostics.* AIAA Infotech@Aerospace Conference, Arlington, VA.

Schwabacher, M., & Goebel, K. (2007). *A Survey of Artificial Intelligence for Prognostics.* AAAI Fall Symposium, Arlington, VA.

Tang, L., Kacprzynski, G. J., Goebel, K., & Vachtsevanos, G. (2009). *Methodologies for Uncertainty Management in Prognostics.* IEEE Aerospace Conference, Big Sky, MT.

Uckun, S., Goebel, K., & Lucas, P. J. F. (2008). *Standardizing Research Methods for Prognostics.* International Conference on Prognostics and Health Management (PHM08), Denver, CO.

Wang, T., & Lee, J. (2009). *On Performance Evaluation of Prognostics Algorithms.* Machinery Failure Prevention Technology, Dayton, OH.

Wheeler, K. R., Kurtoglu, T., & Poll, S. (2009). *A Survey of Health Management User Objectives Related to Diagnostic and Prognostic Metrics.* ASME 2009 International Design Engineering Technical Conferences and Computers and Information in Engineering Conference (IDETC/CIE), San Diego, CA.

Yang, C., & Letourneau, S. (2007). *Model Evaluation for Prognostics: Estimating Cost Saving for the End Users.* Sixth International Conference on Machine Learning and Applications.

Abhinav Saxena is a Research Scientist with SGT Inc. at the Prognostics Center of Excellence NASA Ames Research Center, Moffett Field CA. His research focuses on developing and evaluating prognostic algorithms for engineering systems. He received a PhD in Electrical and Computer Engineering from Georgia Institute of Technology, Atlanta. He earned his B.Tech in 2001 from Indian Institute of Technology (IIT) Delhi, and Masters Degree in 2003 from Georgia Tech. Abhinav has been a GM manufacturing scholar and is also a member of IEEE, AAAI and AIAA.

Jose R. Celaya is a staff scientist with SGT Inc. at the Prognostics Center of Excellence, NASA Ames Research Center. He received a Ph.D. degree in Decision Sciences and Engineering Systems in 2008, a M. E. degree in Operations Research and Statistics in 2008, a M. S. degree in Electrical Engineering in 2003, all from Rensselaer Polytechnic Institute, Troy New York; and a B.S. in Cybernetics Engineering in 2001 from CETYS University, Mexico.

Bhaskar Saha is a Research Scientist with Mission Critical Technologies at the Prognostics Center of Excellence NASA Ames Research Center. His research is focused on applying various classification, regression and state estimation techniques for predicting remaining useful life of systems and their components. He completed his PhD from the School of Electrical and Computer Engineering at Georgia Institute of Technology in 2008. He received his MS from the same school and his B. Tech. (Bachelor of Technology) degree from the Department of Electrical Engineering, Indian Institute of Technology, Kharagpur.

Sankalita Saha received her B.Tech (Bachelor of Technology) degree in Electronics and Electrical Communication Engineering from Indian Institute of Technology, Kharagpur, India in 2002 and Ph.D. in Electrical and Computer Engineering from University of Maryland, College Park in 2007. She is currently a Research scientist with Mission Critical Technologies at NASA Ames Research Center, Moffett Field, CA. Her research interests are in prognostics algorithms and architectures, distributed systems, and system synthesis.

Kai Goebel received the degree of Diplom-Ingenieur from the Technische Universität München, Germany in 1990. He received the M.S. and Ph.D. from the University of California at Berkeley in 1993 and 1996, respectively. Dr. Goebel is a senior scientist at NASA Ames Research Center where he leads the Diagnostics & Prognostics groups in the Intelligent Systems division. In addition, he directs the Prognostics Center of Excellence and he is the Associate Principal Investigator for Prognostics of NASA's Integrated Vehicle Health Management Program. He worked at General Electric's Corporate Research Center in Niskayuna, NY from 1997 to 2006 as a senior research scientist. He has carried out applied research in the areas of artificial intelligence, soft computing, and information fusion. His research interest lies in advancing these techniques for real time monitoring, diagnostics, and prognostics. He holds eleven patents and has published more than 100 papers in the area of systems health management.

Empirical Evaluation of Diagnostic Algorithm Performance Using a Generic Framework

Alexander Feldman[1], Tolga Kurtoglu[2], Sriram Narasimhan[3], Scott Poll[4], David Garcia[5], Johan de Kleer[6], Lukas Kuhn[6], Arjan van Gemund[1]

[1] *Delft University of Technology, Delft, 2628 CD, The Netherlands*
{a.b.feldman,a.j.c.vangemund}@tudelft.nl
[2] *Mission Critical Technologies @ NASA Ames Research Center, Moffett Field, CA, 94035, USA*
tolga.kurtoglu@nasa.gov
[3] *University of California, Santa Cruz @ NASA Ames Research Center, Moffett Field, CA, 94035, USA*
sriram.narasimhan-1@nasa.gov
[4] *NASA Ames Research Center, Moffett Field, CA, 94035, USA*
scott.poll@nasa.gov
[5] *Stinger Ghaffarian Technologies @ NASA Ames Research Center, Moffett Field, CA, 94035, USA*
david.garcia@nasa.gov
[6] *Palo Alto Research Center, 3333 Coyote Hill Road, Palo Alto, CA 94304, USA*
{lukas.kuhn,dekleer}@parc.com

ABSTRACT

A variety of rule-based, model-based and data-driven techniques have been proposed for detection and isolation of faults in physical systems. However, there have been few efforts to comparatively analyze the performance of these approaches on the same system under identical conditions. One reason for this was the lack of a standard framework to perform this comparison. In this paper we introduce a framework, called DXF, that provides a common language to represent the system description, sensor data and the fault diagnosis results; a run-time architecture to execute the diagnosis algorithms under identical conditions and collect the diagnosis results; and an evaluation component that can compute performance metrics from the diagnosis results to compare the algorithms. We have used DXF to perform an empirical evaluation of 13 diagnostic algorithms on a hardware testbed (ADAPT) at NASA Ames Research Center and on a set of synthetic circuits typically used as benchmarks in the model-based diagnosis community. Based on these empirical data we analyze the performance of each algorithm and suggest directions for future development.

1 INTRODUCTION

Fault Diagnosis in physical systems involves the detection of anomalous system behavior and the identification of its cause. Some key steps in diagnostic inference are fault detection (is the output of the system incorrect?), fault isolation (what is

This is an open-access article distributed under the terms of the Creative Commons Attribution 3.0 United States License, which permits unrestricted use, distribution, and reproduction in any medium, provided the original author and source are credited.

Submitted 2/2010; published 7/2010.

broken in the system?), fault identification (what is the magnitude of the failure?), and fault recovery (how can the system continue to operate in the presence of the faults?). To develop diagnostic inference algorithms requires expert knowledge and prior know-how about the system, models describing the behavior of the system, and operational sensor data. This problem is challenging for a variety of reasons including:

- incorrect and/or insufficient knowledge about system behavior

- limited observability

- presence of many types of faults (such as system, supervisor, actuator, or sensor faults; additive and multiplicative faults; abrupt and incipient faults; persistent and intermittent faults; etc.)

- non-local and delayed effect of faults due to dynamic nature of the system.

- presence of other phenomena that influence/mask the symptoms of faults (unknown inputs acting on system, noise that affects the output of sensors, etc.)

Several communities have attempted to solve the diagnostic inference problem using various methods. Some approaches have been:

- Expert Systems - These approaches encode knowledge about system behavior into a form that can be used for inference. Some examples are rule-based systems (Russell & Norvig, 2003) and fault trees (Kavčič & Juričić, 1997).

- Model-Based Methods - These approaches use an explicit model of the system configuration and behavior to guide the diagnostic inference. Some examples are Fault Detection and Isolation (FDI) methods (Gertler, 1998), statistical methods (Basseville & Nikiforov, 1993), and "Artificial Intelligence (AI)" methods (Reiter, 1987).

- Data-Driven Methods - These approaches use the data from representative runs to learn parameters that can then be used for anomaly detection or diagnostic inference for future runs. Some examples are Inductive Monitoring System (IMS) (Iverson, 2004), and Neural Networks (Sorsa & Koivo, 1998).

- Stochastic Methods - These approaches treat diagnosis as a belief state estimation problem. Some examples are Bayesian Networks (Lerner, Parr, Koleer, & Biswas, 2000), and Particle Filters (de Freitas, 2002).

Despite the development of such a variety of notations, techniques, and algorithms, efforts to evaluate and compare diagnostic algorithms (DAs) have been minimal. One of the major deterrents is the lack of a common framework for evaluating and comparing diagnostic algorithms. The establishment of such a framework would accomplish the following objectives:

- Accelerate research in theories, principles, modeling and computational techniques for diagnosis of physical systems.

- Encourage the development of software platforms that promise more rapid, accessible, and effective maturation of diagnostic technologies.

- Provide a forum for algorithm developers to test and validate their technologies.

- Systematically evaluate diagnostic technologies by producing comparable performance assessments.

Such a framework requires the following:

- A standard representation format for the system description, sensor data, and diagnosis result.

- A software run-time architecture that can run specific scenarios from actual system, simulation, or other data sources such as files (individually or as a batch), execute DAs, send scenario data to the DA at appropriate time steps, and archive the diagnostic results from the DA.

- A set of metrics to be computed based on the comparison of the actual scenario and diagnosis results from the DA.

In this paper, we present a framework that attempts to address each of the above issues. The framework architecture employed for evaluating the performance of DAs is shown in Fig. 1 and is called DXF. Major elements are systems under diagnosis, DAs, scenario-based experiments, and metrics. System catalogs specify topology, components, and high-level mode behavior descriptions, including failure modes. DXF provides a program for quantitatively evaluating the DA output against known fault injections using predefined metrics.

The current version of DXF and this paper address a class of abrupt failures such as the ones often observed in electrical power systems. Other types of failures, for example intermittent or continuous ones, are left for future work.

1.1 Contributions

The contributions of this paper are as follows:

- It introduces a benchmarking framework to be used for systematic empirical evaluation of diagnostic algorithm performance. Moreover, it defines and describes the main elements of the framework so that the benchmarking results can be applied to any arbitrary physical or synthetic system by using the architecture described in the paper.

- It provides a comprehensive set of empirical evaluation results in order to validate the proposed framework and to facilitate the understanding and comparative analysis of different diagnostic technologies.

1.2 Organization of the Paper

The rest of this paper is organized as follows. Section 2 contains related work. Section 3 presents DXF in detail including the representation languages used, the run-time architecture developed for experimentation, and the diagnostic performance metrics defined. Section 4 describes how the benchmarking was performed including a description of the two systems used, the faults injected, the DAs tested, and the results. Section 5 presents major assumptions made and issues observed. Finally, Section 6 presents the conclusions.

2 RELATED WORK

The development of monitoring and diagnostic technologies is of great interest to many applications. As these algorithms become more readily available, the necessity for assessing the performance of alternative diagnostic tools becomes important. As a result, there is an increasing need for a framework to evaluate of competing diagnostic technologies.

To address this need, several researchers have attempted to demonstrate benchmarking capability (Orsagh, Roemer, Savage, & Lebold, 2002; Roemer, Dzakowic, Orsagh, Byington, & Vachtsevanos, 2005; Bartyś, Patton, Syfert, de las Heras, & Quevedo, 2006). Among these, Bartyś et al. (2006) presented a benchmarking study for actuator fault detection and identification (FDI). This study, developed by the DAMADICS Research Training Network, introduced a set of 18 performance indices used for benchmarking FDI algorithms on an industrial valve-actuator system. The indices measure the temporal performance of detection and isolation decisions, as well as true and false detection and isolation rates, sensitivity, and diagnostic accuracy. This benchmark study uses real process data, and demonstrates how the performance indices can be calculated for 19 actuator faults using a single fault assumption.

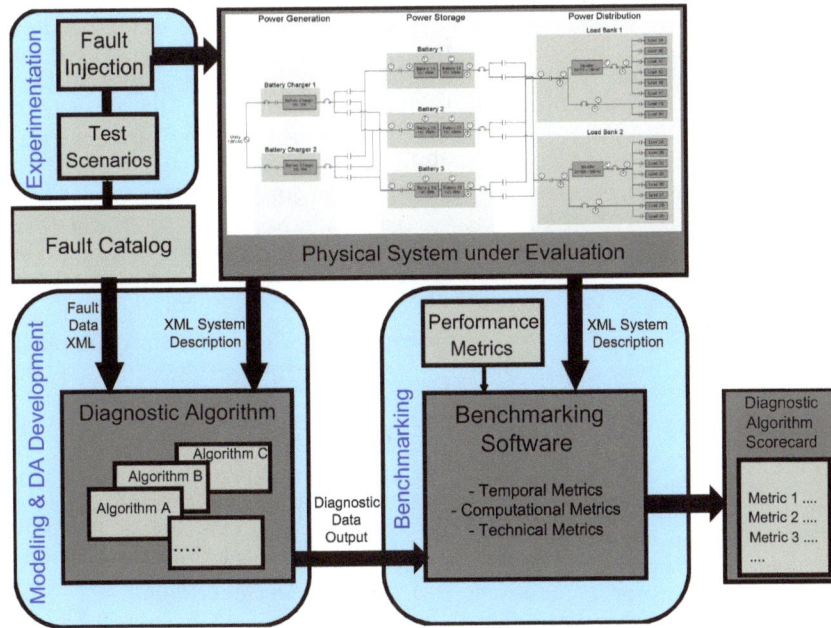

Figure 1: Framework architecture

Izadi-Zamanabadi and Blanke (1999) presented a ship propulsion system as a benchmark for autonomous fault control. This benchmark has two main elements. One is the development of an FDI algorithm, and the other is the analysis and implementation of autonomous fault accommodation.

Relevant to aerospace industry, Simon, Bird, Davison, Volponi, and Iverson (2008) introduced a benchmarking technique for gas path diagnosis methods to assess the performance of engine health management technologies.

Finally, Orsagh et al. (2002) provided a method to measure the performance and effectiveness of prognostics and health management algorithms for US Navy applications (Roemer et al., 2005). In this work, the performance metrics are defined separately for detection, isolation, and prognosis. In addition, this work also combined individual metrics into a composite score by implementing a weighted average sum. Moreover, it defined effectiveness metrics as a separate category that can be used to incorporate non-technical aspects such as operation, maintenance and implementation costs, computer resource requirements, and algorithm complexity into the analysis. Using these metrics, one can assess the overall effectiveness and benefit of diagnostic health management systems.

Other researchers have also proposed similar cost-benefit formulations for diagnostic systems (Williams, 2006; Kurien & Moreno, 2008; Hoyle, Mehr, Tumer, & Chen, 2007). These approaches, however, are primarily concerned with higher-level trade-offs in integrating diagnostic solutions to provide health management functionality and focus on performance indices such as operational cost, and maintainability.

The DXF framework presented in this paper

adopts some of its metrics from Kurtoglu, Narasimhan, Poll, Garcia, Kuhn, de Kleer, van Gemund, and Feldman (2009) and extends prior work in this area by defining a number of novel diagnostic performance metrics; by providing a generic, application independent architecture that can be used for evaluating different monitoring and diagnostic algorithms; and by facilitating the use of real process data on a large-scale, complex engineering system.

3 FRAMEWORK

We have developed a framework called DXF that allows systematic comparison and evaluation of diagnostic algorithms under identical experimental conditions. The key components of this framework include representation languages for the physical system description, sensor data and diagnosis results, a runtime architecture for executing diagnostic algorithms and diagnostic scenarios, and an evaluation component that computes performance metrics based on the results from diagnostic algorithm execution.

The process to set up the framework in order to perform comparison/evaluation of a selected set of diagnostic algorithms on a specific physical system is as follows:

1. The system is specified in an XML file called the System Catalog. The catalog includes the system's components, connections, components' operating modes, and a textual description of component behavior in each mode.

2. The set of sensor points is chosen and sample data for nominal and fault scenarios are generated.

3. DA developers use the system catalog and sample data to create their algorithms using a predefined Application Programming Interface (API) in order to receive sensor data and send the diagnosis results. The DXF API is described later in this section.

4. A set of test scenarios (nominal and faulty) is selected to evaluate the DAs.

5. The run-time architecture is used to execute the DAs on the selected test scenarios in a controlled experiment setting, and the diagnosis results are archived.

6. Selected metrics are computed by comparing actual scenarios and diagnosis results from DAs. These metrics are then used to compute secondary metrics.

In the following subsections we describe the constituent pieces of our framework in more detail. The next subsection describes the various representation languages defined for the framework. We then describe the run-time architecture including the sequence of events and the messages exchanged among the various components and finally we describe a set of representative metrics that measure diagnostic performance.

3.1 DXF Data Structures

In what follows we describe the syntax and semantics of the relevant DXF data structures as well as some design rationale.

3.1.1 System Description

We realize that it is impossible to avoid bias towards certain diagnostic algorithms and methodologies when providing system descriptions. Despite attempts to create a general modeling language[1], there is no widely agreed way to represent models and systems. On the other hand, designing a diagnostic framework which is fully agnostic towards the system description is impossible as there would be no way to communicate components or system parts and to compute diagnostic metrics. As a compromise, we have chosen a minimalistic approach, providing formal descriptions of the system topology and component modes only.

The formal part of the DXF system description does not provide all information for building a model. The user may be provided with non-formalized external information, e.g., nominal and faulty functionality of components. This information may be provided in textual, programmatic or any other well-understood format. In the future we may try to extend our XML schema in yet another attempt of providing a complete modeling language beyond interconnection topology.

The XML system description is primarily intended to provide a common set of identifiers for components and their modes of operation within a given system. This is necessary to communicate sensor data and diagnoses. Additionally, basic structural information is provided in the form

---

[1]For examples see Feldman, Provan, and van Gemund (2007) and the references therein.

of component connections. Behavioral information is limited to a brief textual description of each component and its modes, leaving DA developers to deduce behavior from the system's sample data. This is done to avoid bias towards any diagnostic approach.

System Topology: DXF uses a graph-like representation to specify the physical connectivity of the system where nodes represent components of a system and arcs capture the connectivity between components.

Component Types: Each component in a system description refers to a component type. Note that in DXF, sensors do not imply special assumptions, i.e., sensors fail in the same way as "ordinary" components. A sensor, of course, should specify the data type it returns in order for DXF to send sensor readings to the DA under evaluation. A component type contains at least the following information:

- a name (identifier)
- an optional (textual) description
- a flag which specifies if this component type is a sensor
- a reference to a data structure describing the modes for the components of this type (both nominal and faulty)
- (sensors only) a data type of the sensor
- (sensors only) a range of the sensor

Component Mode Groups: Component operating modes are organized in mode groups. More than one component can refer to the same specific group. Each component type specifies a mode group. Each mode in a mode group contains:

- a name (identifier)
- an optional (textual) description
- a flag specifying if the mode is nominal or faulty

The details of the system description formats are provided in Appendix B.

3.1.2 API Data Types

In DXF, the run-time communication is performed using a messaging framework. Messages are exchanged as ASCII text over TCP/IP. API calls for parsing, sending, and receiving messages are provided with the framework, but developers may choose to send and receive messages directly through the underlying TCP/IP interface. This allows developers to use their programming language of choice, rather than being forced into the languages of the provided APIs.

Every message contains a millisecond timestamp indicating the time at which the message was sent. Though there are additional message types, the most important messages for the purpose of performance evaluation are the sensor data message, command message, and diagnosis message, described below (the details of the messaging formats are provided in Appendix C):

Sensor/Command Data: Sensor data are defined broadly as a map of sensor IDs to sensor values (observations). Sensor values can be of any type; currently the framework allows for integer, real, Boolean, and string values. The type of each observation is indicated by the system's XML catalog.

Commandable components contain an additional entry in the system catalog specifying a command ID and command value type (analogous to sensor value type). The command message represents the issuance of a command to the system. In the ADAPT system, for example, the message (EY144CL, true) signifies that relay EY144 is being commanded to close. EY144CL is the command ID, and true is the command value (in this case, a Boolean value).

Candidates: The diagnostic algorithm's output (i.e., estimate of the physical status of the system) is standardized to facilitate the generation of common data sets and the calculation of the performance metrics. The diagnostic message contains:

- a timestamp value indicating when the diagnosis has been issued by the algorithm
- a list of diagnostic candidates (a candidate fault set may include a single candidate with a single or multiple faults; or multiple candidates each with a single or multiple faults)
- a detection flag (Boolean) as to whether the diagnosis system has detected a fault
- an isolation flag (Boolean) as to whether the diagnosis system has isolated a candidate or a set of candidates

In addition, each candidate in the candidate set has an associated weight. Candidate weights are normalized by DXF such that their sum for any given diagnosis is 1.

3.2 Run-Time Architecture

Figure 2 shows an overview of the DXF run-time architecture, its software components and data flows.

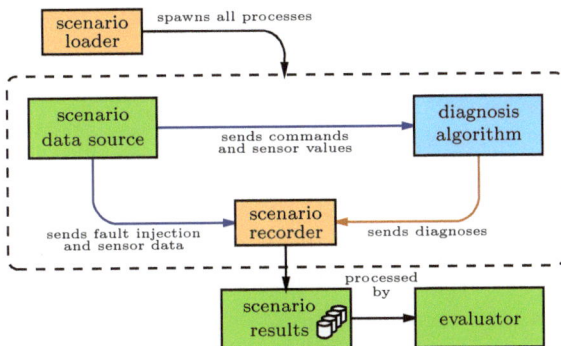

Figure 2: DXF run-time architecture

We next provide a brief description of each of the DXF's software components.

Scenario Loader (SL): SL is the main entry point for running the diagnostic scenarios. SL executes the Scenario Data Source, the Scenario Recorder, and all Diagnostic Algorithms. SL ensures system stability and clean-up upon scenario completion and is the only long-living process. The Scenario Data Source, Scenario Recorder and all Diagnostic Algorithms are spawned for each scenario and a Diagnostic Algorithm is forcibly killed if it does not terminate after a predetermined time-out.

Scenario Data Source (SDS): The SDS module provides scenario data from previously recorded datasets. The provenance of the data (whether hardware or simulation) depends on the system in question. A scenario dataset contains sensor readings, commands (note that the majority of classical MBD literature does not distinguish commands from observations), and fault injection information (to be sent exclusively to SR). SDS publishes data following a wall-clock schedule specified by timestamps in the scenario files.

Scenario Recorder (SR): SR receives fault injection data and diagnosis data into a results file. The results file contains a number of time-series which are described below. These time-series are used by the evaluation module for the computation of metrics. SR is the main timing authority, i.e., it timestamps each message upon arrival before recording it to the results file.

Diagnostic Algorithm (DA): A DA receives sensor and command data, performs diagnosis, and sends the diagnosis results back. As long as the DAs comply to the provided API, there are no restrictions on a DA; for example a DA may read precompiled data, or use external (user supplied) libraries, etc.

Evaluator: The evaluator computes a number of predefined metrics (see Sec. 3.3).

Consider the progression of a single diagnostic scenario. A typical one is shown in Fig. 3, where the fault injections, detection, and isolation are all treated as signals. These signals define a number of time points and intervals, as is seen below.

In the beginning of each scenario, a DA is given some startup time to initialize, read data, etc. Even though sensor observations could be available during startup, fault injections are not allowed during this interval. Fault injection and diagnosis take place during the diagnosis interval. Finally, a DA is given some shutdown time to terminate before being killed.

Table 1 summarizes the data collected by the SR for each scenario. These data are used for computing the various metrics discussed in Sec. 3.3. The time of first detection t_d is derived from the detec-

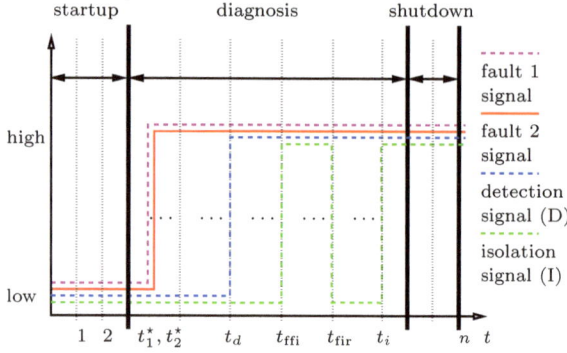

Figure 3: Signals and events during the development of a sample diagnostic scenario

tion signal D while the time of the last isolation is computed from the isolation signal I.

Var.	Type	Description	Origin
t_d	time-stamp	first detection	DA
t_i	time-stamp	last isolation	DA
C_s	real	startup CPU cycles	SL
C	time-series	CPU cycles per step	SL
M	time-series	memory in use	SL
$\omega^\star$	set	injected fault	SDS
$t^\star$	time-stamp	first fault injection	SDS
$t_i^\star$	time-stamp	injection of fault i	SDS
Ω	set of sets	candidate diagnoses	DA
W	set of reals	candidate weights	DA

Table 1: Scenario execution summary data

The set $\Omega = \{\omega_1, \omega_2, \ldots, \omega_n\}$ contains all diagnoses computed by the DA at time t_i. If a DA never asserts the isolation signal I (i.e., $t_i = \infty$), it is assumed that $\Omega = \emptyset$. Each candidate in Ω is accompanied by a weight W. We denote the set of weights of all diagnoses in Ω as $W = \{W(\omega_1), W(\omega_2), \ldots, W(\omega_n)\}$. The SR ensures that

$$\sum_{\omega \in \Omega} W(\omega) = 1 \qquad (1)$$

by dividing each weight $W(\omega)$ with the sum of all weights. If a DA fails to provide W, it is assumed that all diagnoses are of the same weight.

In addition to the time-points defined in Table 1, the isolation signal in Fig. 3 shows the time t_{ffi} the DA has isolated a fault for the first time, and the time t_{fir} the DA has retracted its isolation assumption (for example because more faults are expected). Note that t_{ffi} and t_{fir} are not currently used by the evaluator for computing the metrics.

3.3 Diagnostic Performance Metrics

The metrics for evaluating diagnostic algorithm performance depend on the particular use of the diagnostic system, the users involved, and their objectives.

Several institutions and organizations have proposed metrics that measure diagnostic performance (Committee E-32, 2008; DePold, Siegel, & Hull, 2004; DePold, Rajamani, Morrison, & Pattipati, 2006; Metz, 1978; Orsagh et al., 2002; Roemer et al., 2005; Bartyś et al., 2006). Among those, the SAE's "Health and Usage Monitoring Metrics" (Committee E-32, 2008) defines probability of detection and probability of false alarms as key indices for evaluating diagnostic algorithm performance.

In Orsagh et al. (2002), the performance metrics are defined separately for detection, and isolation. For detection, the metrics include thresholds, accuracy, reliability, sensitivity to load, speed, or noise, and stability. The isolation metrics include the detection metrics, but also include measures for discrimination and repeatability.

In this paper, our goal has been to define a number of metrics and to give guidelines for their use. For DXF, we make a distinction between detection, isolation, and computational performance and highlight metrics for each category. In general several other classes of metrics are possible, including cost/utility metrics, effort (in building systems for example) metrics and also other categories such as fault identification and fault recovery metrics. The expectation is that as the DXF evolves a comprehensive list of desired metric classes and categories will be developed to aid framework users in choosing the performance criteria they want to measure.

Metric	Name	Class
M_{fd}	fault detection time	detection
M_{fn}	false negative scenario	detection
M_{fp}	false positive scenario	detection
M_{da}	scenario detection accuracy	detection
M_{fi}	fault isolation time	isolation
M_{err}	classification errors	isolation
M_{utl}	utility	isolation
M_{sat}	consistency	isolation
M_{cpu}	CPU load	computational
M_{mem}	memory load	computational

Table 2: Metrics summary

For the first implementation of the DXF, we defined 10 metrics which are summarized in Table 2. These metrics are based on extensive survey of literature and talking to experts from various fields (Kurtoglu, Mengshoel, & Poll, 2008). These metrics are defined next.

3.3.1 Detection Metrics

The distinction between detection and isolation has practical importance. A DA may announce a fault detection before it knows the root cause of failure (for example, a detection announcement can be based solely on surpassing sensor threshold values). A detection signal cannot be retracted by a DA while it is legal to retract an isolation announcement when more faults are expected. The detection metrics include:

Fault Detection Time The *fault detection time* (the reaction time for a diagnostic engine to detect an anomaly) is directly measured as:

$$M_{\text{fd}} = t_d \tag{2}$$

The fault detection time is reported in milliseconds and is computed only for non-nominal scenarios for which a DA asserts the time detection signal at least once.

False Negative Scenario The *false negative scenario* metric measures whether a fault is missed by a diagnostic algorithm and is defined as:

$$M_{\text{fn}} = \begin{cases} 1, & \text{if } t_d = \infty \\ 0, & \text{otherwise} \end{cases} \tag{3}$$

False Positive Scenario The *false positive scenario* metric penalizes DAs which announce spurious faults and is defined as:

$$M_{\text{fp}} = \begin{cases} 1, & \text{if } t_d < t^\star \\ 0, & \text{otherwise} \end{cases} \tag{4}$$

where $t^\star = \infty$ for nominal scenarios (i.e., scenarios during which no fault is injected).

Note that the above two metrics (M_{fn} and M_{fp}) are computed for each scenario and their computation is based on the times of injecting and announcing the fault. We also have false negative and false positive components in the context of individual diagnostic candidates (recall that a DA sends a set of diagnostic candidates at isolation time) which we will discuss later in this paper.

Scenario Detection Accuracy The *scenario detection accuracy* metric is computed from M_{fn} and M_{fp}:

$$M_{\text{da}} = 1 - \max(M_{\text{fn}}, M_{\text{fp}}) \tag{5}$$

M_{da} is 1 if the scenario is true positive or true negative and 0 otherwise (equivalently, $M_{\text{da}} = 0$ if $M_{\text{fn}} = 1$ or $M_{\text{fp}} = 1$, and $M_{\text{da}} = 1$ otherwise). M_{da} splits all scenarios into "true" and "false". Incorrect scenarios are further classified into false positive (M_{fp}) and false negative (M_{fn}). Correct scenarios are true positive if there are injected faults and true negative otherwise (the latter separation into true positives and true negatives is rarely of practical importance).

3.3.2 Isolation Metrics

Computation of isolation metrics is more involved due to the fact that an isolation can be retracted. Furthermore, an isolation event contains a set of diagnostic candidates and we need metrics that compare this set of candidates to the injected fault. Accordingly, we have defined several metrics which are computed from the set of diagnostic

candidates Ω and the injected fault $\omega^\star$ (classification errors, and utility metrics). Consider a single diagnostic candidate $\omega \in \Omega$. Both the candidate ω and the injected fault $\omega^\star$ are sets of components. The intersection of those two sets are the properly diagnosed components. The false positives are the components that have been considered faulty but are not actually faulty. The false negatives are the components that have been considered healthy but are actually faulty. Figure 4 shows how ω and $\omega^\star$ partition all components into four sets.

False positives and false negatives in this context relate to individual candidates, i.e., misclassified components in a single diagnostic candidate. There are also scenario-based false negative and false positive metrics (defined earlier in this section), which summarize whole scenarios and are not to be confused with the false positives and false negatives in the context of isolation metrics.

For brevity we use the notation in Table 3 for the Fig. 4 sets.

Var.	Set	Description
f	|COMPS|	all components
n	$\|\omega^\star \setminus \omega\|$	false negatives
N	|COMPS $\setminus \omega$|	the set of healthy components from the viewpoint of the DA
$\bar{n}$	$\|\omega \setminus \omega^\star\|$	false positives
$\bar{N}$	$\|\omega\|$	the set of faulty components from the viewpoint of the DA

Table 3: Notation for sizes of some frequently used sets

Based on the representation given in Figure 4, the meaning of false positives and false negatives can be interpreted differently depending on what the diagnosis results are supporting (abort decisions, ground support, fault-adaptive control, etc.). Researchers have proposed different methods to assess the meaning of isolation accuracy and its practical and economical implications. DePold et al. (2004) introduced metrics based on the receiving operating characteristic (ROC) analysis (Metz, 1978), which illustrates the trade-off space between the probability of false alarm and the probability of detection for different signal to noise ratio (SNR) levels. The method is used to test the relative accuracy of diagnostic systems based on different threshold settings. Later, they also proposed a combined metric (DePold et al., 2006) that accounts for consequential event costs including missed detection, false alarms, and misdiagnosis. Another widely used metric for isolation accuracy is the Kappa Coefficient (Committee E-32, 2008). It is based on the construction of a confusion matrix that summarizes diagnostic results produced by a reasoner over a number of test/use cases. In essence, the Kappa Coefficient measures the ability of an algorithm to discriminate among many fault candidates.

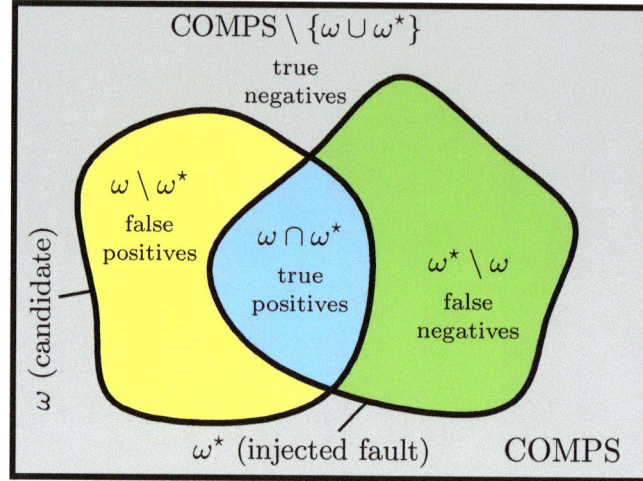

Figure 4: The diagnostic candidate ω and the injected fault $\omega^\star$ partition COMPS into four sets

In this paper, we take a simplistic approach and assume that false positives and false negatives have an equal cost for the diagnostic task and operations. The isolation metrics include (for a detailed discussion and derivation of the isolation metrics, see Appendix A):

Fault Isolation Time Consider an injected fault $\omega^\star = \{c_1, c_2, \ldots, c_n\}$ with the individual component faults injected at times $T^\star = \langle t_1^\star, t_2^\star, \ldots, t_n^\star \rangle$. Next, from the isolation signal, we construct a sequence of isolation times for each component. This sequence containing timestamps of rising edges of the isolation signal is denoted as T_i ($T_i = \langle t_1, t_2, \ldots, t_n \rangle$). Note that $t_k^\star < t_i$ for $1 \le k \le n$. The *fault isolation time* is then computed as:

$$M_{\text{fi}} = \frac{1}{n} \sum_{k=1}^{n} t_k - t_k^\star \tag{6}$$

If there is no isolation for specific fault (i.e., a fault is missed) then there is no difference $t_k - t_k^\star$ computed for that fault. E.g., if in a fault $\omega^\star = \langle c_1, c_2, c_3 \rangle$, c_1 is isolated, c_2 is not, and c_3 is; the isolation time $t_2 - t_2^\star$ is undefined and not included in the average ($n = 2$).

The *fault isolation time* is reported in milliseconds and is computed only for non-nominal scenarios for which a DA asserts the time isolation signal at least once.

Classification Error The *classification error* metric is defined as:

$$M_{\text{err}} = \sum_{\omega \in \Omega} W(\omega)(|\omega \ominus \omega^\star|) \tag{7}$$

In Eq. (7), $\omega \ominus \omega^\star$ denotes the symmetric difference of the ω and $\omega^\star$ sets, i.e., the number of misclassified components. Note that $|\omega \ominus \omega^\star| = n + \bar{n}$ and $f = N + \bar{N}$.

Utility The *utility* metric measures the work for correctly identifying all false negatives and false positives in a diagnostic candidate. Alternatively, the utility metric measures the expected number of calls to a testing oracle that always determines correctly the health state of a component. Note that this metric assumes an equal cost for fixing a false negative and a false positive. The derivation of the utility metric is given in Appendix A. The utility metric (per candidate) is:

$$m_{\text{utl}} = 1 - \frac{n(N+1)}{f(n+1)} - \frac{\bar{n}(\bar{N}+1)}{f(\bar{n}+1)} \tag{8}$$

Computing a weighted average of m_{utl} gives us the "per scenario" utility metric:

$$M_{\text{utl}} = \sum_{\omega \in \Omega} W(\omega) m_{\text{utl}}(\omega^\star, \omega) \tag{9}$$

The utility metric is, in fact, a combination of two "half-utilities"–the system repair utility and the diagnosis repair utility. The latter are defined as secondary metrics in Sec. 3.3.4 and discussed in detail in Appendix A.

Note that for $\Omega = \emptyset$, the framework automatically assumes a single "all-healthy" diagnostic candidate with weight 1 at the time of isolation. This affects the M_{err} and M_{utl} metrics. For example, in a non-nominal false-negative scenario, $M_{\text{err}} = |\omega^\star|$.

Consistency The next metric comes from MBD (de Kleer, Mackworth, & Reiter, 1992). It only applies to systems for which (1) there is a formally defined system description (model), (2) one can derive a formally defined observation from the sensor data, and (3) the notion of consistency is formally defined. We compute the consistency metric for the synthetic models and scenarios.

Consider a model SD and an observation α (α is derived from the sensor data at time t^*). If SD and α can be expressed as sentences in propositional logic (as is the case with the synthetic models and scenarios) then the set of consistent diagnoses is defined as:

$$\Omega^\top = \{\omega \in \Omega : \text{SD} \wedge \alpha \wedge \omega \not\models \bot\} \qquad (10)$$

The set $\Omega^\top$ can be computed from SD, α, and Ω by using a DPLL-solver (Davis, Logemann, & Loveland, 1962). The *consistency* metric can be computed from $\Omega^\top$, W and the injected fault $\omega^\star$:

$$M_{\text{sat}} = \sum_{\omega \in \Omega^\top} W(\omega) \qquad (11)$$

M_{sat} is a measure of how much probability mass a DA associates with diagnoses consistent with the observations.

3.3.3 Computational Metrics

CPU Load The *CPU load* during an experiment is computed as:

$$M_{\text{cpu}} = C_s + \sum_{c \in C} c \qquad (12)$$

where C_s is the amount of CPU time spent by a DA during startup and C is a vector with the actual CPU time spent by the DA at each time step. The CPU load is reported in milliseconds.

Memory Load The *memory load* is defined as:

$$M_{\text{mem}} = \max_{m \in M} m \qquad (13)$$

where M is a vector with the maximum memory size allocated at each step of the diagnostic session. The memory load is reported in Kb.

3.3.4 Secondary Metrics

The intuition behind classification errors can be realized with multiple metrics. For example, a diagnostician may compute the isolation accuracy using:

$$M_{\text{ia}} = \sum_{\omega \in \Omega} W(\omega)(f - |\omega \ominus \omega^\star|) \qquad (14)$$

In general a diagnostician has to perform extra work to "verify" all misdiagnosed components in ω. Suppose that the diagnostician has access to a test oracle that states if a component c is healthy or faulty. The system repair utility is then defined as normalized average number of oracle calls for identifying all false negative components and is defined as:

$$m_{\text{sru}} = 1 - \frac{n(N+1)}{f(n+1)} \qquad (15)$$

The "per scenario" system repair utility is the defined as:

$$M_{\text{sru}} = \sum_{\omega \in \Omega} W(\omega)m_{\text{sru}}(\omega^\star, \omega) \qquad (16)$$

Similarly, a diagnostician has to eliminate all false positive components in a candidate. This is reflected in the diagnosis repair utility:

$$m_{\text{dru}} = 1 - \frac{\bar{n}(\bar{N}+1)}{f(\bar{n}+1)} \qquad (17)$$

The diagnosis repair utility for a set of diagnostic candidates is defined as:

$$M_{\text{dru}} = \sum_{\omega \in \Omega} W(\omega)m_{\text{dru}}(\omega^\star, \omega) \qquad (18)$$

M_{utl}, M_{sru}, and M_{dru} are discussed in detail in Appendix A.

The choice of which utility metric is best for a particular use depends on the relative costs of the available repair actions. For example, if components are nearly free, but the act of replacing them is expensive then it makes no sense to identify which erroneously replaced components were actually correct (thus m_{sru} is preferred).

3.3.5 System Metrics

The metrics M_{fn}, M_{fp}, M_{da}, M_{fd}, M_{fi}, M_{err}, M_{utl}, M_{sat}, M_{cpu}, M_{mem}, M_{ia}, M_{sru}, and M_{dru} are based on a single scenario. To receive "per system" results we combine the metrics of each scenario using unweighted average. For example, if a system SD is tested with scenarios $\mathbf{S} = \{S_1, S_2, \ldots, S_n\}$, the "per system" utility of SD is computed as:

$$\bar{M}_{\text{utl}} = \sum_{S \in \mathbf{S}} \frac{1}{|\mathbf{S}|} M_{\text{utl}}(\text{SD}, S) \qquad (19)$$

where $M_{\text{utl}}(\text{SD}, S)$ is the "per scenario" utility of system SD and scenario S.

The rest of the "per system" metrics ($\bar{M}_{\text{fn}}$, $\bar{M}_{\text{fp}}$, $\bar{M}_{\text{da}}$, $\bar{M}_{\text{fd}}$, $\bar{M}_{\text{fi}}$, $\bar{M}_{\text{err}}$, $\bar{M}_{\text{sat}}$, $\bar{M}_{\text{cpu}}$, $\bar{M}_{\text{mem}}$, $\bar{M}_{\text{ia}}$, $\bar{M}_{\text{sru}}$, and $\bar{M}_{\text{dru}}$) are defined in a way analogous to $\bar{M}_{\text{utl}}$.

Note that M_{fn}, M_{fp}, and M_{da} are called false negative scenario, false positive scenario and scenario detection accuracy, respectively. The analogous "per system" metrics $\bar{M}_{\text{fn}}$, $\bar{M}_{\text{fp}}$, and $\bar{M}_{\text{da}}$ are called *false negative rate*, *false positive rate*, and *detection accuracy*. $\bar{M}_{\text{da}}$, for example, represents the ratio of the number of correctly classified cases to the total number of cases. The latter "per system" metrics ($\bar{M}_{\text{fn}}$, $\bar{M}_{\text{fp}}$, and $\bar{M}_{\text{da}}$) are equivalent to the ones in Kurtoglu et al. (2009). In this paper we first define each metric "per scenario" and then "per system".

4 EMPIRICAL EVALUATION

In order to evaluate the framework presented in the previous section we selected two case studies. The first case study was performed on an Electrical Power System (EPS) testbed located in the ADAPT Lab of NASA Ames Research Center (Poll, Patterson-Hine, Camisa, Garcia, Hall, Lee, Mengshoel, Neukom, Nishikawa, Ossenfort, Sweet, Yentus, Roychoudhury, Daigle, Biswas, & Koutsoukos, 2007). This system mimics components and configurations in a power system that might be found on an aerospace vehicle. The second case study was performed on a set of 14 synthetic systems called the 74XXX/ISCAS85 circuits (Brglez & Fujiwara, 1985), which are purely combinational, i.e., they contain no flip-flops or other memory elements, and represent well-known benchmark models of ISCAS85 circuits.

The empirical evaluation as part of the above two case studies employed 13 diagnostic algorithms (DAs) (Kurtoglu et al., 2009). The results from the DAs were used to compute metrics that were in turn used to evaluate the DAs performance on the aforementioned systems. We first present the DAs used in the evaluation and then present the two case studies.

4.1 Diagnostic Algorithms

We have experimented with a total of 13 DAs (see Table 4 for an overview). In what follows we provide a brief description of each DA.

DA	Systems	Algorithm Type
FACT	AL	model-based
Fault Buster	A,AL	statistical
GoalArt	A	flow-models
HyDE	A,AL	model-based
HyDE-S	AL	model-based
LYDIA	S,A,AL	model-based
NGDE	S,AL	model-based
ProADAPT	A,AL	probabilistic
RacerX	AL	change detection
RODON	S,A,AL	model-based
RulesRule	AL	rule-based
StanfordDA	A	optimization
Wizards of Oz	A,AL	model-based

Table 4: Diagnostic Algorithms (S = synthetic, A = ADAPT, AL = ADAPT-Lite)

FACT: FACT (Roychoudhury, Biswas, & Koutsoukos, 2009) is a model-based diagnosis system that uses hybrid bond graphs, and models derived from them, at all levels of diagnosis, including fault detection, isolation, and identification. Faults are detected using an observer-based approach with statistical techniques for robust detection. Faults are isolated by matching qualitative deviations caused by fault transients to those predicted by the model. For systems with few operating configurations, fault isolation is implemented in a compiled form to improve performance.

Fault Buster: Fault Buster is based on a combination of multivariate statistical methods, for the generation of residuals. Once the detection has been done a neural network performs classification for computing isolation.

GoalArt: GoalArt Diagnostic System (Larsson, 1996) is based on multilevel flow models, which are crisp descriptions of flows of mass, energy, and information. It is a fast root cause analysis with linear computational complexity. Its main advantage is that it is very efficient to knowledge engineer a model. The algorithm has been proven in several commercial applications.

HyDE: HyDE (Hybrid Diagnosis Engine) (Narasimhan & Brownston, 2007) is a model-based diagnosis engine that uses consistency between model predictions and observations to generate conflicts which in turn drive the search for new fault candidates. HyDE uses discrete models of the system and a discretization of the sensor observations for diagnosis.

HyDE-S: HyDE-S uses the HyDE system but runs it on interval valued hybrid models and the raw sensor data.

Lydia: LYDIA is a declarative modeling language specifically developed for Model-Based Diagnosis (MBD). The language core is propositional logic, enhanced with a number of syntactic extensions for ease of modeling. The accompanying toolset currently comprises a number of diagnostic engines and a simulator tool (Feldman, Provan, & van Gemund, 2009).

NGDE: An Allegro Common Lisp implementation of the classic GDE. NGDE (de Kleer, 2009) uses a minimum-cardinality candidate generator to construct diagnoses from conflicts. For ADAPT-Lite it uses interval constraints. No model of dynamics.

ProADAPT: ProADAPT (Mengshoel, 2007) processes all incoming environment data (observations from a system being diagnosed), and acts as a gateway to a probabilistic inference engine. The inference engine uses an Arithmetic Circuit evaluator which is compiled from Bayesian network models. The primary advantage of using arithmetic circuits is speed, which is key in resource bounded environments.

RacerX: RacerX is a detection-only algorithm which detects a percentage change in individual filtered sensor values to raise a fault detection flag.

RODON: RODON (Karin, Lunde, & Münker, 2006) is based on the principles of the General Diagnostic Engine (GDE) as described by de Kleer and Williams (1987) and the G$^+$DE (Heller & Struss, 2001). RODON uses contradictions (conflicts) between the simulated and the observed behavior to generate hypotheses about possible causes for the observed behavior. If the model contains failure modes in addition to the nominal behavior, these can be used to verify the hypotheses, which speeds up the diagnostic process and improves the results.

RulesRule: RulesRule is a rule-based isolation-only algorithm. The rule base was developed by analyzing the sample data and determining characteristic features of faults. There is no explicit fault detection though isolation implicitly means that a fault has been detected.

StanfordDA: StanfordDA is an optimization-based approach to estimating fault states in

DC power systems. The model includes faults changing the system topology along with sensor faults. The approach can be considered as a relaxation of the mixed estimation problem. The authors have developed a linear model of the circuit and pose a convex problem for estimating the faults and other hidden states. A sparse fault vector solution is computed by using L1 regularization (Zymnis, Boyd, & Gorinevsky, 2009).

Wizards of Oz: Wizards of Oz (Grastien & Kan-John, 2009) is a consistency-based algorithm. The model of the system completely defines the stable (static) output of the system in case of normal and faulty behavior. Given a new command or new observations, the algorithm waits for a stable state and computes the minimum diagnoses consistent with the observations and the previous diagnoses.

4.2 Case Study I: ADAPT EPS

We next describe the ADAPT EPS system, the diagnostic scenarios and the experimental results.

4.2.1 System Description

The ADAPT EPS testbed provides a means for evaluating DAs through the controlled insertion of faults in repeatable failure scenarios. The EPS testbed incorporates low-cost commercial off-the-shelf (COTS) components connected in a system topology that provides the functions typical of aerospace vehicle electrical power systems: energy conversion/generation (battery chargers), energy storage (three sets of lead-acid batteries), power distribution (two inverters, several relays, circuit breakers, and loads) and power management (command, control, and data acquisition).

The EPS delivers Alternating Current (AC) and Direct Current (DC) power to loads, which in an aerospace vehicle could include subsystems such as the avionics, propulsion, life support, environmental controls, and science payloads. A data acquisition and control system commands the testbed into different configurations and records data from sensors that measure system variables such as voltages, currents, temperatures, and switch positions. Data are presently acquired at a 2 Hz rate.

The scope of the ADAPT EPS testbed used in this case study is shown Fig. 5. Power storage and distribution elements from the batteries to the loads are within scope; there are no power generation elements defined in the system catalog. We have created two systems from the same physical testbed, ADAPT-Lite and ADAPT:

ADAPT-Lite ADAPT-Lite includes a single battery and a single load as indicated by the dashed lines in the schematic (Fig. 5). The initial configuration for ADAPT-Lite data has all relays and circuit breakers closed and no nominal mode changes are commanded during the scenarios. Hence, any noticeable changes in sensor values may be correctly attributed to faults injected

into the scenarios. Furthermore, ADAPT-Lite is restricted to single faults.

ADAPT ADAPT includes all batteries and loads in the EPS. The initial configuration for ADAPT has all relays open and nominal mode changes are commanded during the scenarios. The commanded configuration changes result in adjustments to sensor values as well as transients which are nominal and not indicative of injected faults, in contrast to ADAPT-Lite. Finally, multiple faults may be injected in ADAPT. The differences between ADAPT-Lite and ADAPT are summarized in Table 5.

Aspect	ADAPT-Lite	ADAPT
|COMPS|	37	173
# of modes	93	430
relays initially	closed	open
initial state of the circuit-breakers	closed	closed
nominal mode changes	no	yes
multiple faults	no	yes

Table 5: ADAPT and ADAPT-Lite differences

4.2.2 Diagnostic Challenges

The ADAPT EPS testbed offers a number of challenges to DAs. It is a hybrid system with multiple modes of operation due to switching elements such as relays and circuit breakers. There are continuous dynamics within the operating modes and components from multiple physical domains, including electrical, mechanical, and hydraulic. It is possible to inject multiple faults into the system. Furthermore, timing considerations and transient behavior must be taken into account when designing DAs. For example, when power is input to the inverter there is a delay of a few seconds before power is available at the output. For some loads, there is a large current transient when the device is turned on. System voltages and currents depend on the loads attached, and noise in sensor data increases as more loads are activated. Measurement noise occasionally exhibits spikes and is non-Gaussian. The 2 Hz sample rate limits the types of features that may be extracted from measurements. Finally, there may be insufficient information and data to estimate parameters of dynamic models in certain modeling paradigms.

4.2.3 Fault Injection and Scenarios

ADAPT supports the repeatable injection of faults into the system in three ways:

Hardware-Induced Faults: These faults are physically injected at the testbed hardware. A simple example is tripping a circuit breaker using the manual throw bars. Another is using the power toggle switch to turn off an inverter. Faults may also be introduced in the loads attached to the EPS. For example, the valve can be closed slightly to vary the

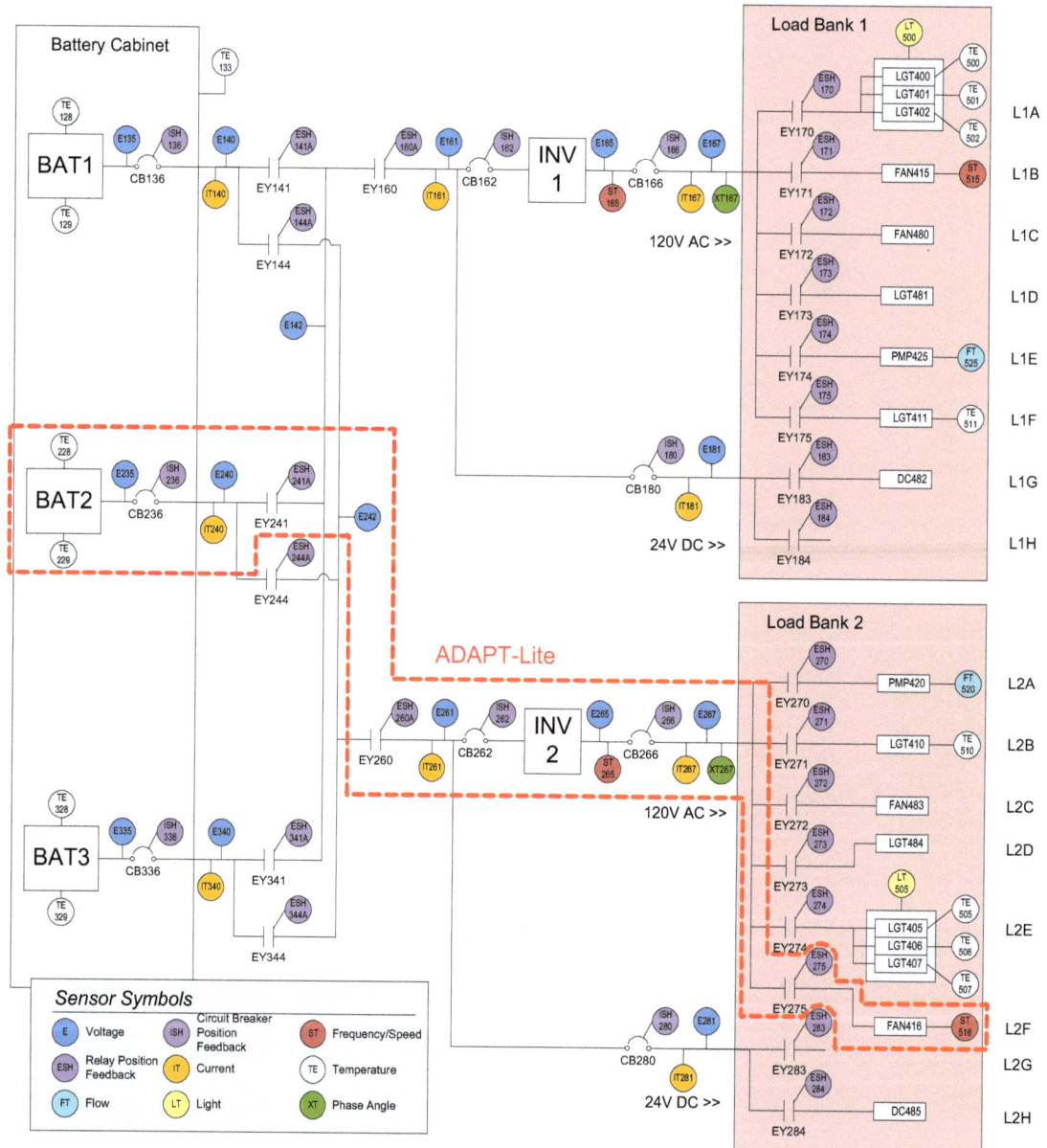

Figure 5: A schematic overview of the ADAPT EPS

back pressure on the pump and reduce the flow rate.

Software-Induced Faults: In addition to fault injection through hardware, faults may be introduced via software. Software fault injection includes one or more of the following: (1) sending commands to the testbed that are not intended for nominal operations; (2) blocking commands sent to the testbed; and (3) altering the testbed sensor data.

Real Faults: In addition the aforementioned two methods, real faults may be injected into the system by using actual faulty components. A simple example includes a burned out light bulb. This method of fault injection was not used in this study.

For results presented in this case study, only abrupt discrete (change in operating mode of component) and parametric (step change in parameter value) faults are considered. Nominal and failure scenarios are created using hardware and software-induced fault injection methods. The diagnostic algorithms are tested against a number of scenarios, each approximately four minutes in length.

The ADAPT-Lite experiments include 36 nominal and 56 single-fault scenarios. Table 6 summarizes the type of faults used for ADAPT-Lite.

Type	Subtype	Fault	#
battery	-	degraded	3
circ. breaker	-	failed-open	5
inverter	-	failed-off	2
load	fan	failed-off	2
		over-speed	2
		under-speed	2
relay	-	stuck-open	6
sensor	position	stuck	11
	current, phase angle, speed, temp., voltage	offset	12
		stuck	11
		Total:	56

Table 6: ADAPT-Lite faults

The ADAPT experiments have 48 nominal and 111 fault scenarios, which include single-fault, double-fault, and triple-faults. Figure 6 shows the fault-cardinality distribution of the ADAPT scenarios. Table 7 summarizes the type of faults used for ADAPT. The majority of faults involve sensors (102) and loads (30).

4.2.4 Experimental Results
We next compute the metrics described in Sec. 3.3 for the ADAPT-Lite and ADAPT scenarios.

ADAPT-Lite The DA benchmarking results for ADAPT-Lite are shown in Table 8, with graph-

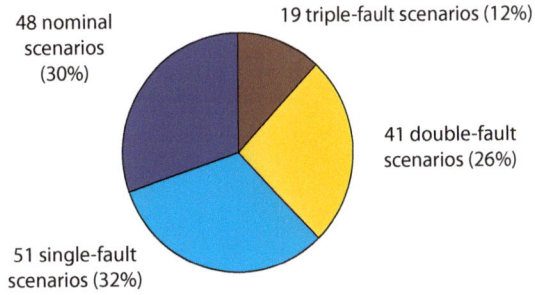

Figure 6: Fault-cardinality distribution of the ADAPT scenarios

Type	Subtype	Fault	#
battery	-	degraded	1
circuit breaker	-	failed-open	18
inverter	-	failed-off	10
load	basic	failed-off	1
	fan	failed-off	5
		over-speed	2
		under-speed	3
	light bulb	failed-off	14
	pump	failed-off	3
		blocked	2
relay		stuck-closed	3
		stuck-open	26
sensor	position	stuck	26
	current, flow, light, phase angle, speed, temp., voltage	offset	35
		stuck	41
		Total:	190

Table 7: ADAPT faults

ical depictions of some of the tabular data presented in Fig. 7. Figure 7 shows (1) M_{err} by DA (top-left), (2) M_{sru} and M_{dru} by DA (top-right), (3) M_{fd} and M_{fi} by DA (bottom-left), and (4) M_{fn} and M_{fp} (bottom-right). No DA dominates over all metrics used in benchmarking; nine of eleven DAs tested are best or second best with respect to at least one of the metrics.

The bottom-right plot of Fig. 7 shows the false positive and false negative rates. The corresponding detection accuracy can be seen in Table 8. As is evident from the definition of the metrics in Sec. 3.3, a DA that has low false positive and negative rates has high detection accuracy. False positives are counted in the following two situations: (1) for nominal scenarios where the DA declares a fault; and (2) for faulty scenarios where the DA declares a fault before any fault is injected. Noise in the data and incorrect models are the main causes

DA	Detection				Isolation			Computation	
	$\bar{M}_{\text{fd}}$	$\bar{M}_{\text{fn}}$	$\bar{M}_{\text{fp}}$	$\bar{M}_{\text{da}}$	$\bar{M}_{\text{fi}}$	$\bar{M}_{\text{err}}$	$\bar{M}_{\text{utl}}$	$\bar{M}_{\text{cpu}}$	$\bar{M}_{\text{mem}}$
FACT	1 785	0	0.11	0.89	10 798	11	0.975	15 815	4 271
Fault Buster	155	0.5	0.01	0.68	–	56	0.685	1 951	2 569
HyDE	13 355	0.46	0	0.72	13 841	45	0.79	23 418	5 511
HyDE-S	121	0.04	0.38	0.6	683	66	0.791	573	5 366
Lydia	232	0.18	0.01	0.88	232	100.3	0.785	1 410	1 861
NGDE	194	0.13	0.03	0.89	14 922	44.5	0.833	21 937	73 031
ProADAPT	4 732	0.05	0.01	0.96	7 104	10	0.955	1 905	1 226
RacerX	77	0.2	0.03	0.85	–	56	0.685	146	3 619
RODON	4 204	0.04	0.01	0.97	12 364	4	0.983	12 050	28 870
RulesRule	949	0.09	0.33	0.62	949	63	0.818	167	3 784
Wizards of Oz	12 202	0.5	0	0.7	12 327	43	0.769	1 153	1 682

Table 8: ADAPT-Lite metrics results

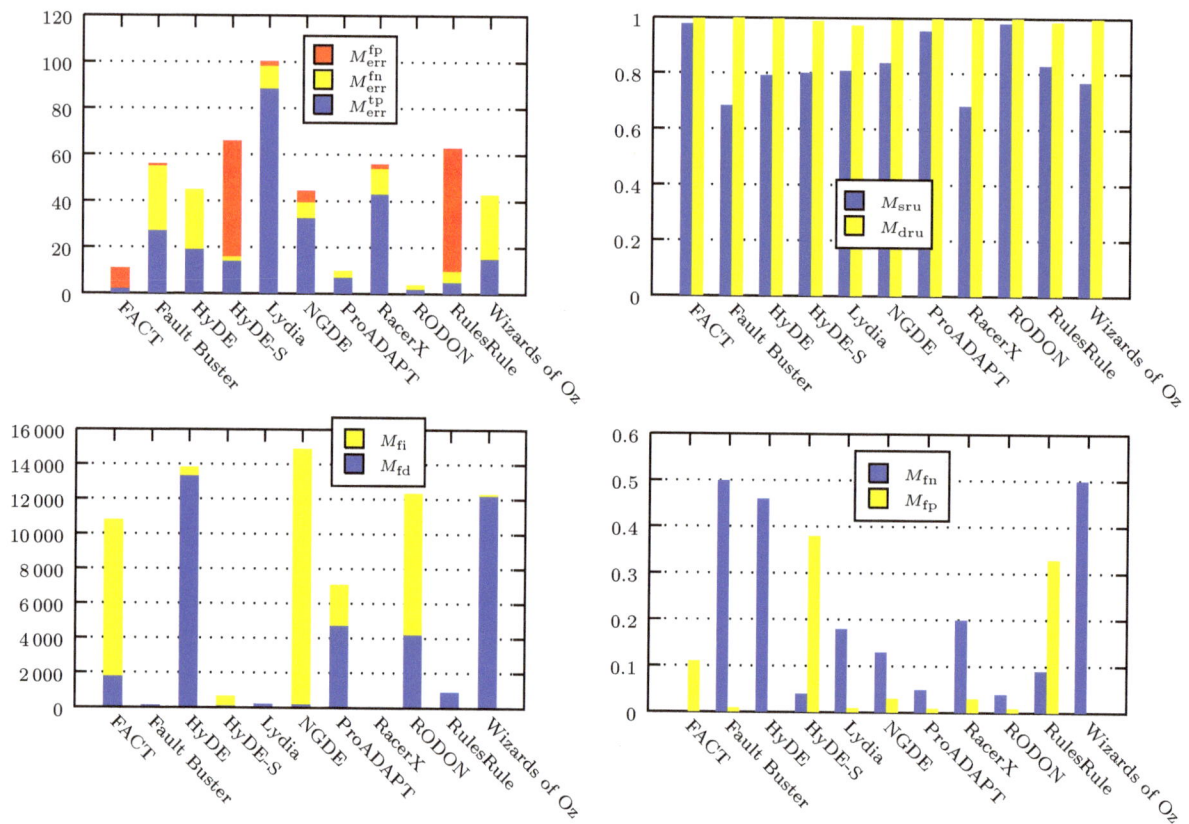

Figure 7: ADAPT-Lite metrics

of false positives. For example, the leftmost plot of Fig. 8 shows a nominal run with spike in sensor IT240 (battery 2 current); most of the DAs declare a false positive for this scenario. Many false negatives are caused by scenarios in which a sensor reading is stuck within the nominal range of the sensor. The middle plot of Fig. 8 shows an example of a sensor-stuck failure for voltage sensor E261, the downstream voltage of relay EY260.

The classification error metric for each DA is shown in the top-left plot of Fig. 7, where the error contributions of scenarios labeled false neg-

ative, false positive, and true positive are noted. Many DAs have difficulties distinguishing between sensor-stuck and sensor-offset faults. The distinction in the fault behavior is that stuck has zero noise while offset has the noise of the original signal; the rightmost plot in Fig. 8 shows the fan speed sensor ST516 with sensor-offset and sensor-stuck faults. In many scenarios, the sensor-stuck faults are set to the minimum or maximum value of the sensor or held at its last reading. The latter case presents the most difficulties to DAs.

M_{fd} and M_{fi} are shown in the bottom-left plot

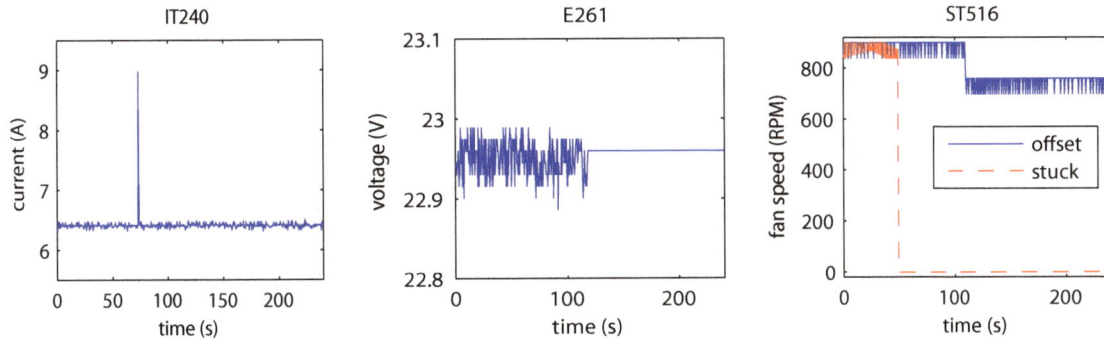

Figure 8: Examples of sensor readings

of Fig. 7. RacerX is a detection-only DA and does not perform isolation (its detection time is very low). Note that $M_{\mathrm{fd}} \leq M_{\mathrm{fi}}$, hence the bottom-left plot of Fig. 7 shows the isolation time stacked on the detection time (assume that part of the time first goes into detection and then into isolation).

The top-right plot of Fig. 7 shows the system repair utilty, M_{sru}, and the diagnosis repair utility, M_{dru}. The diagnosis repair utility is very close to 1 for all DAs, which reflects the small fault cardinality and diagnosis ambiguity groups for the system. The number of components that a DA considers faulty, $\bar{N}$, in any given scenario is typically close to the number of faults injected in the scenario. Since $\bar{N}$ is much less than the number of components, f, it is evident from equation (17) that M_{dru} approaches 1. Furthermore, since the number of healthy components, N, as determined by the DA is larger than the number of faulty components, $\bar{N}$, whereas n is typically not much different from $\bar{n}$, the system repair utility is smaller than the diagnosis repair utility.

Note that HyDE has been used by two different modelers of ADAPT-Lite. HyDE was modeled primarily with the larger and more complex ADAPT in mind and had a policy of waiting for transients to settle before requesting a diagnosis. The same policy was applied to ADAPT-Lite as well, even though transients in ADAPT-Lite corresponded strictly to fault events; this prevented false positives in ADAPT but negatively impacted the timing metric in ADAPT-Lite. On the other hand, HyDE-S was modeled only for ADAPT-Lite and did not include a lengthy time-out period for transients to settle. HyDE-S had dramatically smaller mean detection and isolation times (see the bottom-left plot of Fig. 7) with roughly the same M_{err} (see Table 8) as HyDE. This illustrates the impact that modeling and implementation decisions have on DA performance. While this gives some insight into trade-offs present in building models, in this work we did not define metrics that directly address the ease or difficulty of building models of sufficient fidelity for the diagnosis task at hand.

As is visible from Table 8, there exist significant differences in M_{cpu} and M_{mem}. Part of these differences can be attributed to the operating system (Linux or Windows™). RODON was the only Java DA that was run on Windows™, which adversely affected its memory usage metric.

ADAPT The empirical DA benchmarking results for ADAPT are shown in Table 9. Figure 9 shows (1) M_{err} by DA (top-left), (2) M_{sru} and M_{dru} by DA (top-right), (3) M_{fd} and M_{fi} by DA (bottom-left), and (4) M_{fn} and M_{fp} (bottom-right). Five of eight DAs tested were best or second best with respect to at least one of the metrics for ADAPT.

The comments in the ADAPT-Lite discussion about noise and sensor stuck apply here as well. Additionally, false positives also result from nominal commanded mode changes in which the relay feedback did not change status as of the next data sample after the command. Here is an extract from one of the input scenario files that illustrates this situation:

```
command @120950 EY275_CL = false;
sensors @121001 {..., ESH275 = true, ... };
sensors @121501 {..., ESH275 = false, ... };
```

A command is given at 120.95 seconds to open relay `EY275`. The associated relay position sensor does not indicate open as of the next sensor data update 51 milliseconds later. This is nominal behavior for the system. A DA that does not account for this delay will indicate a false positive in this case.

The detection and isolation times are generally within the same order of magnitude for the different DAs (see the bottom-left plot of Fig. 9). Some DAs have isolation times that are similar to its detection times while others show isolation times that are much greater than the detection times. This could reflect differences in reasoning strategies or differences in policies for when to declare an isolation based on accumulated evidence.

The CPU and memory usage are shown in Table 9. The same comment for RODON mentioned previously in regards to memory usage applies here. The convex optimization approach applied in the StanfordDA and the compiled arithmetic circuit in ProADAPT lead to very low CPU usages.

DA	Detection				Isolation			Computation	
	$\bar{M}_{\mathrm{fd}}$	$\bar{M}_{\mathrm{fn}}$	$\bar{M}_{\mathrm{fp}}$	$\bar{M}_{\mathrm{da}}$	$\bar{M}_{\mathrm{fi}}$	$\bar{M}_{\mathrm{err}}$	$\bar{M}_{\mathrm{utl}}$	$\bar{M}_{\mathrm{cpu}}$	$\bar{M}_{\mathrm{mem}}$
Fault Buster	21 255	0.39	0.03	0.70	100 292	193	0.587	10 051	7 119
GoalArt	3 268	0.05	0.03	0.93	7 805	154	0.776	149	6 784
HyDE	15 612	0.31	0	0.79	20 114	174.3	0.668	28 807	19 135
Lydia	16 135	0.2	0.25	0.62	16 135	234.9	0.653	5 715	3 412
ProADAPT	1 743	0.02	0.09	0.90	23 544	57	0.915	4 260	778
RODON	5 543	0.03	0	0.98	35 792	75.6	0.853	85 331	31 459
StanfordDA	3 826	0.05	0.17	0.79	16 816	176.6	0.706	1 012	2 213
Wizards of Oz	25 695	0.09	0.16	0.77	50 980	209.2	0.76	17 111	3 390

Table 9: ADAPT metrics results

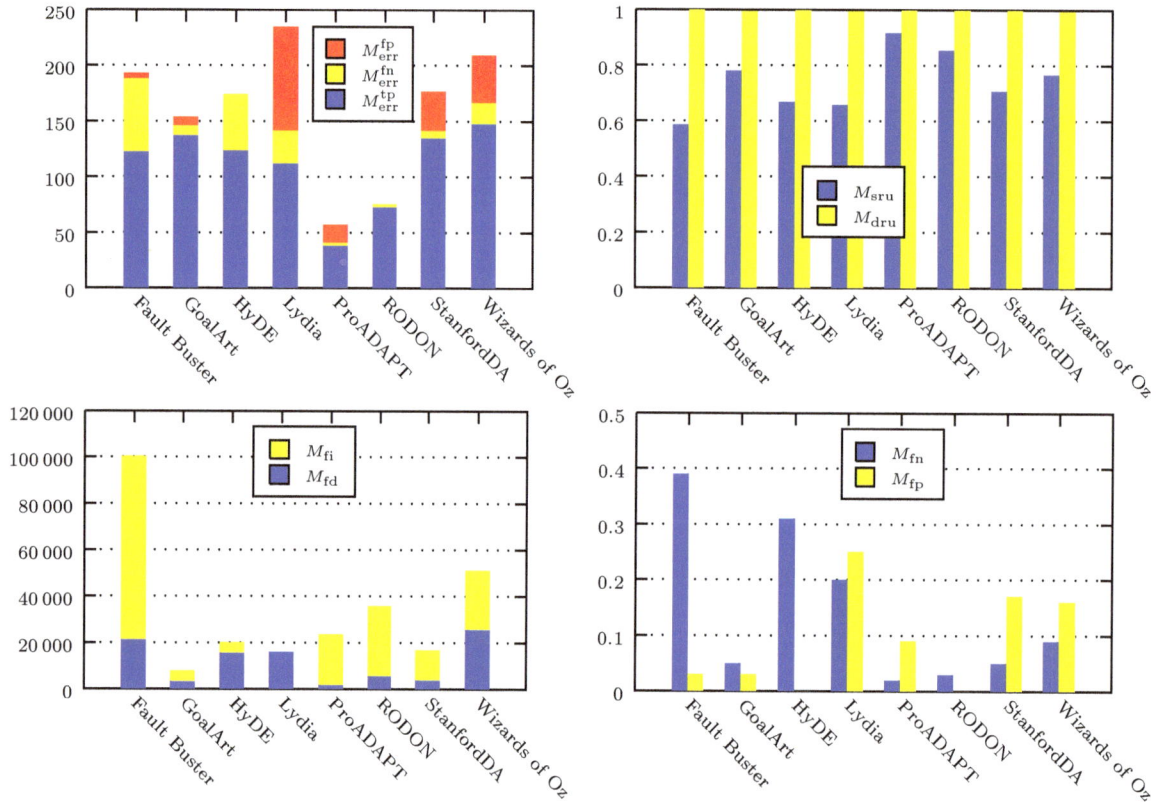

Figure 9: ADAPT metrics

4.2.5 Fault Type and Cardinality Analysis

The plots on the left-hand side of Fig. 10 show detection accuracy for all DAs by fault type for ADAPT-Lite and ADAPT. In general, M_{da} is not very sensitive to the component type, except in the case of load and sensor faults where it is lower. The data on the battery detection accuracy is not representative due to the limited number of fault scenarios containing battery faults (see Table 6 and Table 7).

The plots on the right-hand side of Fig. 10 show classification errors for all DAs by fault type for ADAPT-Lite and ADAPT. While the overall performance (averaged for all DAs) indicates that most fault categories result in roughly the same number of errors per scenario, it can be seen that a given DA may do better on some faults compared to others; furthermore, several DAs have the fewest classification errors for the different fault types. We should also note that in this benchmarking study, no partial credit was given for correctly naming the failed component but incorrectly isolating the failure mode. We realize however, that isolating to a failed component or line-replaceable-unit (LRU) in maintenance operations is sometimes all that is required. We plan to revisit this metric in future work.

Figure 11 shows the breakdown of classification errors by the number of faults in the scenario. In general, the number of errors increased approximately linearly with the number of faults in the

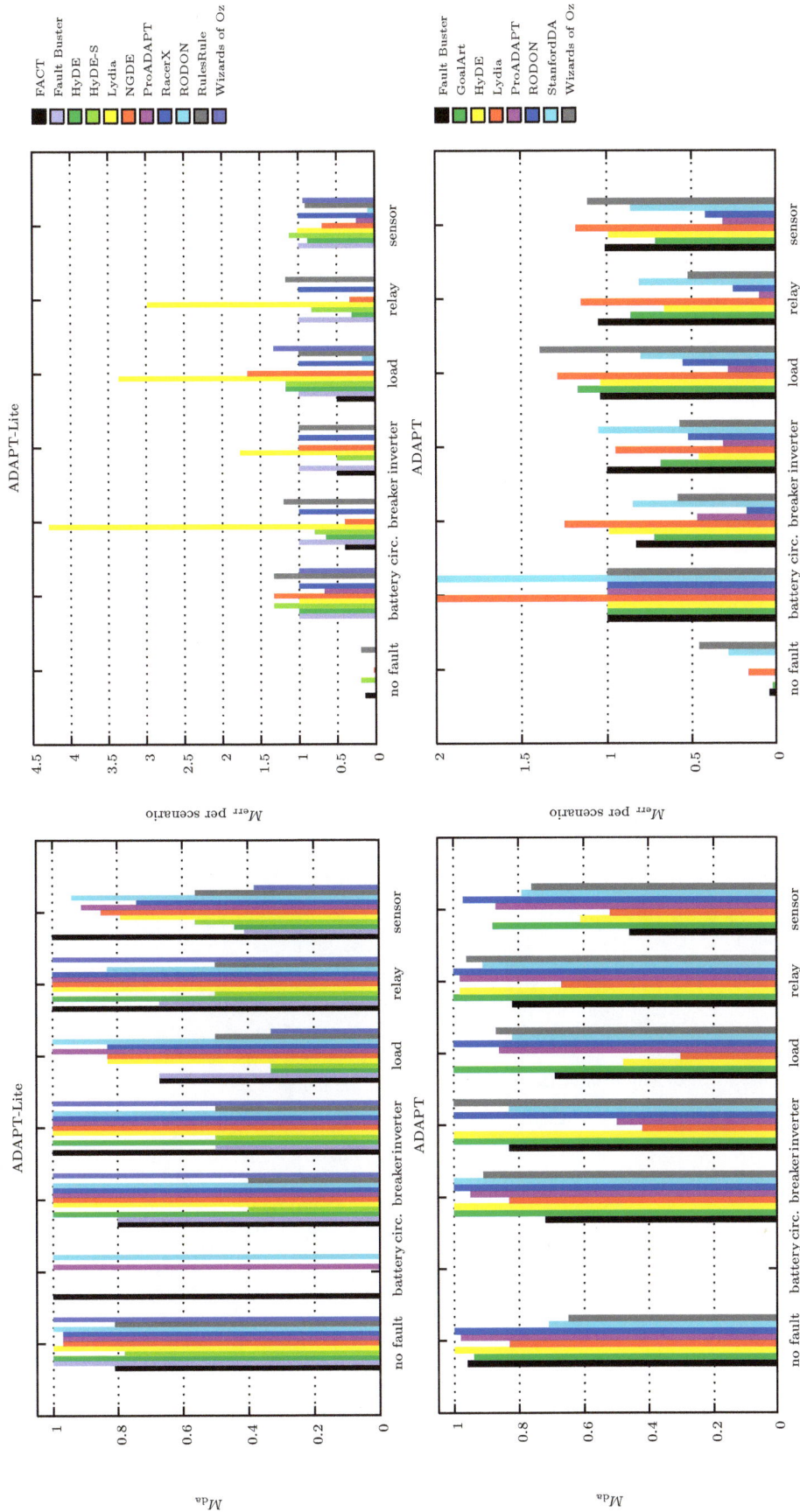

Figure 10: $\bar{M}_{da}$ and $\bar{M}_{err}$ per fault type for all DAs

scenario.The errors in the multiple fault scenarios were evenly divided among the faults; for example, if there were four classification errors in a scenario where two faults were injected, each fault was assigned two errors. We also did a more thorough assessment in which each diagnosis candidate was examined and classification erorrs were assigned to fault categories based on an understanding of which sensors are affected by the faults. The results are similar to evenly dividing the errors among the faults and are not shown here.

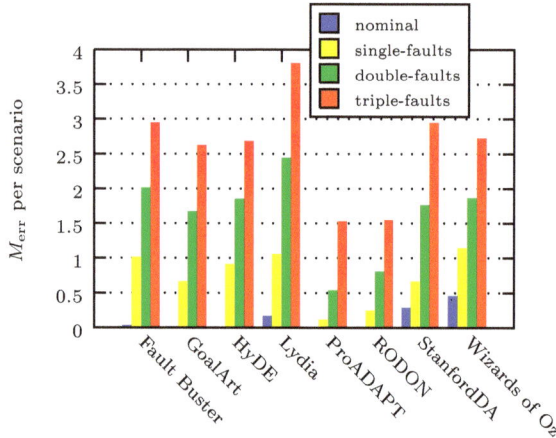

Figure 11: M_{err} per fault cardinality for all DAs (ADAPT)

4.2.6 Metric Correlations

The correlation matrix shown in Table 10 contains the Pearson's linear correlation coefficients between each metric for the industrial systems ADAPT and ADAPT-Lite.

Ideally, metrics should measure different aspects of DAs, i.e., the correlation matrix should contain small values only. Alternatively, users may use the correlation matrix from Table 10 to select metrics and adjust metric weights in computing the parameters of their DAs. Unexpected high correlations (or anti-correlations) between metrics indicate (1) bias due to the system or the sensor data, or (2) hidden metric dependencies.

All correlation coefficients in Table 10, except those shown in bold red, are significant–the p-values according to the Student's t distribution are smaller than 0.03.

Figure 12 is a color map of the correlation matrix from Table 10. Correlations or anticorrelations close to 1 are colored in red, while values closer to 0 are shown in blue colors.

The anti-correlation between M_{ia} and M_{err} is trivial (see (22)) and the only reason for including it is to show the correctness of our implementation.

The utility metric shows high correlation with the isolation accuracy/classification errors ($\rho = 0.75$). This is expected as both metrics measure similar properties of the DAs' results. Less trivial is the high correlation between M_{utl}^- and M_{utl}^+ ($\rho = 0.84$). This indicates that DAs do not show

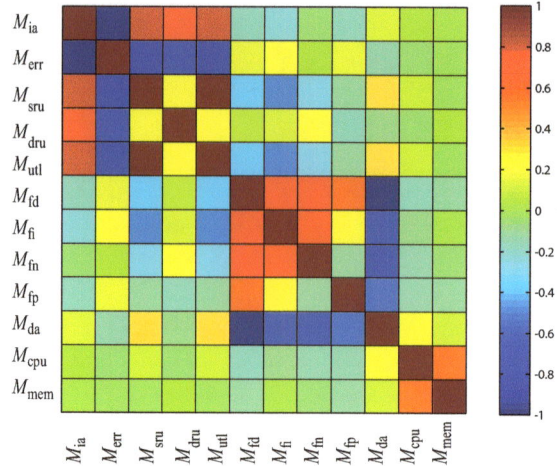

Figure 12: ADAPT & ADAPT-Lite metrics correlations

preferences towards diagnosing false negatives or false positives.

The time for fault isolation M_{ia} correlates highly with the three utility metrics, for which we have no explanation. The M_{fn} metric correlates high with M_{da} which comes from the metric design and indicates that, in general, DAs are tuned to avoid false positives at the price of more false negatives.

4.3 Case Study II: Synthetic Systems

We continue our discussion with an overview of the synthetic systems. The major differences between this case study and the previous are the sizes of the systems and the cardinalities of the injected faults. Furthermore, all system variables in this case study are of Boolean type. This case study aims to compare the robustness, CPU performance, and memory consumption of various DAs under stress conditions (large systems, faults of multiple-cardinality, etc.).

4.3.1 Description of Systems

The original 74XXX/ISCAS85 netlists can be mechanically translated into propositional **Wff**s. We have translated the propositional **Wff**s into logically equivalent Conjunctive Normal Form (CNF) formulae (Forbus & de Kleer, 1993). These CNF formulae are described in Table 11.

For each 74XXX/ISCAS85 CNF formula, Table 11 gives the number of inputs |IN|, the number of outputs |OUT|, the size of the components sets |COMPS|, the number of variables |V|, and the number of clauses |C|. The size of the 74XXX/ISCAS85 circuits can be reduced by using cones for computing single-component ambiguity groups (Siddiqi & Huang, 2007) or using fault collapsing.

4.3.2 Synthetic Model Scenarios

We have noticed that the performance of many DAs depends on the Minimum Cardinality (MC) of the diagnoses. Hence, we have performed our

	M_{ia}	M_{err}	M_{sru}	M_{dru}	M_{utl}	M_{fd}	M_{fi}	M_{fn}	M_{fp}	M_{da}	M_{cpu}	M_{mem}
M_{ia}	1	−1	0.79	0.7	0.8	−0.18	−0.25	**−0.04**	−0.18	0.15	**0.04**	**0.03**
M_{err}	−1	1	−0.79	−0.7	−0.8	0.18	0.25	**0.04**	0.18	−0.15	**−0.04**	**−0.03**
M_{sru}	0.79	−0.79	1	0.21	1	−0.37	−0.5	−0.31	−0.11	0.33	0.1	**−0.01**
M_{dru}	0.7	−0.7	0.21	1	0.24	0.07	0.14	0.23	−0.16	−0.09	**−0.04**	0.06
M_{utl}	0.8	−0.8	1	0.24	1	−0.36	−0.49	−0.3	−0.12	0.33	0.1	**0**
M_{fd}	−0.18	0.18	−0.37	0.07	−0.36	1	0.7	0.75	0.54	−0.98	−0.17	−0.12
M_{fi}	−0.25	0.25	−0.5	0.14	−0.49	0.7	1	0.65	0.19	−0.67	**−0.07**	**0.02**
M_{fn}	**−0.04**	**0.04**	−0.31	0.23	−0.3	0.75	0.65	1	−0.12	−0.76	−0.14	**−0.06**
M_{fp}	−0.18	0.18	−0.11	−0.16	−0.12	0.54	0.19	−0.12	1	−0.55	−0.13	−0.1
M_{da}	0.15	−0.15	0.33	−0.09	0.33	−0.98	−0.67	−0.76	−0.55	1	0.2	0.12
M_{cpu}	**0.04**	**−0.04**	0.1	−0.04	0.1	−0.17	**−0.07**	−0.14	−0.13	0.2	1	0.54
M_{mem}	**0.03**	**−0.03**	**−0.01**	**0.06**	**0**	−0.12	**0.02**	**−0.06**	−0.1	0.12	0.54	1

Table 10: ADAPT metrics correlation matrix (correlations with p-values smaller than 0.03 are shown in bold red)

Name	|IN|	|OUT|	|COMPS|	|V|	|C|
74182	9	5	19	47	75
74L85	11	3	33	77	118
74283	9	5	36	81	122
74181	14	8	65	144	228
c432	36	7	160	356	1 028
c499	41	32	202	445	1 428
c880	60	26	383	826	2 224
c1355	41	32	546	1 133	3 220
c1908	33	25	880	1 793	4 756
c2670	233	140	1 193	2 695	6 538
c3540	50	22	1 669	3 388	9 216
c5315	178	123	2 307	4 792	13 386
c6288	32	32	2 416	4 864	14 432
c7552	207	108	3 512	7 232	19 312

Table 11: 74XXX/ISCAS85 circuits

experimentation with a number of different observations leading to diagnoses of different MCs. Algorithm 1 generates observations leading to diagnoses of different MC, varying from 1 to nearly the maximum for the respective circuits (for the 74XXX models it is the maximum). The experiments omit nominal scenarios as they are trivial with synthetic systems.

The synthetic scenarios disregard the temporal aspects of diagnosis. They are created in the following way. In the beginning of a scenario, a DA is sent a nominal observation. After 5 s a fault $\omega^\star$ is injected. An observation α consistent with $\omega^\star$ is sent 6 s after the scenario start. We next discuss the generation of the "faulty" observations.

Algorithm 1 is an approximate algorithm that returns a set of observations A. Each observation $\alpha \in A$ leads to a diagnosis of different MC and is used in a different scenario. We have executed Alg. 1 multiple times, filtering out identical observations, until we have collected observations for a sufficient number of scenarios.

Algorithm 1 uses a number of auxiliary functions. RandomInputs (line 3) assigns uniformly

Algorithm 1 Algorithm for generation of observation vectors

1: **function** MakeAlphas(DS, N, K) **returns** a set of observations
　　inputs: DS $= \langle$SD, COMPS, OBS$\rangle$
　　　　　OBS $=$ IN $\cup$ OUT, IN $\cap$ OUT $= \emptyset$
　　　　　N, integer, number of tries
　　　　　K, integer, maximal number of
　　　　　　diagnoses per cardinality
　　local variables: $\alpha, \beta, \alpha_n, \omega$, terms
　　　　　c, integer, best card. so far
　　　　　Ω, set of terms, diagnoses
　　　　　A, set of terms, result
2:　　**for** $k \leftarrow 1, 2, \ldots, K$ **do**
3:　　　　$\alpha \leftarrow$ RandomInputs(IN)
4:　　　　$\beta \leftarrow$ NominalOutputs(DS, α)
5:　　　　$c \leftarrow 0$
6:　　　　**for all** $v \in$ OUT **do**
7:　　　　　$\alpha_n \leftarrow \alpha \wedge$ Flip(β, v)
8:　　　　　$\Omega \leftarrow$ Safari(SD, α_n, |COMPS|, N)
9:　　　　　$\omega \leftarrow$ MinCardDiag(Ω)
10:　　　　**if** $|\omega| > c$ **then**
11:　　　　　$c \leftarrow |\omega|$
12:　　　　　$A \leftarrow A \cup \alpha_n$
13:　　　　**end if**
14:　　　**end for**
15:　　**end for**
16:　　**return** A
17: **end function**

distributed random values to each input in IN (note that for the generation of observation vectors we partition the observable variables OBS into inputs IN and outputs OUT and use the input/output information which comes with the original 74XXX/ISCAS85 circuits for simulation). Given the "all healthy" health assignment and the diagnostic system, NominalOutputs (line 4) performs simulation by propagating the input assignment α. The result is an assignment β which contains values for each output variable in OUT.

The loop in lines 7 – 14 increases the cardinality by greedily flipping the values of the output variables. For each new candidate observation

α_n, Alg. 1 uses the diagnostic algorithm SAFARI to compute a minimal diagnosis of cardinality c (Feldman, Provan, & van Gemund, 2008a). As SAFARI returns more than one diagnosis (up to N), we use MINCARDDIAG to choose the one of smallest cardinality. If the cardinality c of this diagnosis increases in comparison to the previous iteration, the observation is added to the list.

By running Alg. 1 we get up to K observations leading to faults of cardinality $1, 2, \ldots, m$, where m is the cardinality of the MFMC diagnosis (Feldman, Provan, & van Gemund, 2008b) for the respective circuit. Alg. 1 clearly shows a bootstrapping problem. In order to create potentially "difficult" observations for a DA we require a DA to solve those "difficult" observations. In our case we have used the anytime SAFARI. As SAFARI is a stochastic algorithm, sometimes it returns a minimal diagnosis when we need a minimal-cardinality one. This leads to scenarios resulting in lower cardinalities than intended but this seemingly causes no problems except minor difficulties in the analysis of the DAs' performance.

4.3.3 Experimental Results

We start this section by computing the relevant metrics for this case study: $\bar{M}_{\text{utl}}$, $\bar{M}_{\text{cpu}}$, and $\bar{M}_{\text{mem}}$. The results are shown in Table 12.

It can be seen that LYDIA has achieved significantly better $\bar{M}_{\text{cpu}}$ and $\bar{M}_{\text{mem}}$ than NGDE and RODON. $\bar{M}_{\text{utl}}$ of LYDIA is slightly worse due to smaller number of diagnostic candidates computed by this DA. LYDIA and RODON showed similar results in the utility metrics.

We have computed $\bar{M}_{\text{sat}}$ and the results are shown in Table 14. The SAT and UNSAT columns show the number of consistent and inconsistent candidates, respectively. NGDE has generated approximately two orders of magnitude more satisfiable candidates than LYDIA and RODON. The policy of LYDIA has been to compute a small number of candidates, minimizing $\bar{M}_{\text{mem}}$ and $\bar{M}_{\text{cpu}}$. In order to improve $\bar{M}_{\text{utl}}$, LYDIA has mapped multiple-cardinality candidates into single-component failure probabilities. Hence, only single-fault scenarios contribute to the $\bar{M}_{\text{sat}}$ score for LYDIA.

5 DISCUSSION

The primary goal of the empirical evaluation presented in this paper was to demonstrate an end-to-end implementation of DXF and create a foundation for future usage of the framework. As a result we made several simplifying assumptions. We also ran into several issues during the course of this implementation that could not be addressed. In this section, we present those assumptions and issues, which we hope can be addressed in future implementations.

5.1 DXF Data Structures

The system catalog has been intentionally defined as a general XML format to avoid committing to specific modeling or knowledge representations (e.g., equations). It is expected that the sample training data and pointers to additional documentation would be sufficient for DA developers to learn the behavior of the system. We will continue to look for ways to extend the system catalog representation to provide as much general information about the system as possible. The diagnosis result format is defined to be a set of candidates with a weight associated with each candidate. Each candidate reports faulty modes of 0 (all nominal) or more components. Obviously this is a simplistic representation since it does not allow reporting of intermittent faults, parametric faults, among others. Also, in some cases it may be desirable to report a belief state (a probability distribution over component states) as opposed to a set of candidates.

5.2 Run-Time Architecture

For the ADAPT system, the fault signatures were limited to abrupt parametric and discrete types. We plan to introduce other fault types (incipient, intermittent, and noise) in the future. The runtime architecture was defined such that no assumptions were made regarding the actual operational environments in which the diagnostic algorithms may be run. We understand that a true test would simulate operating conditions of the real system, i.e. the system operates nominally for long periods of time and failures occur periodically following the prior probability of failure distribution. In this work, faults were inserted assuming equal probabilities. In the future, we will provide the failure rates of components and use these to evaluate the performance of DAs. It was also assumed that all sensor data was available to the DAs at all time steps. In the future, we would like to relax this assumption and provide only a subset of the sensor data. Additional ideas for future research include giving DAs reduced sensor sets, introducing multi-rate sensor data, injecting transient faults, allowing for autonomous transitions, adding variable loads, and extending the scope and complexity of the physical system.

For the synthetic systems, all the systems have been known in advance. This means researchers could optimize for these circuits. In addition, only one observation time was sampled. In the future, we will provide multiple observations. This will evaluate a DA's ability to merge information from multiple times. An important component of troubleshooting is introducing probe points. In the future, we can evaluate the number of probes needed to isolate the fault.

5.3 Diagnostic Metrics

The set of metrics we have chosen as primary is based on literature survey and expert opinion on what measures are important to assess the effectiveness of DAs. However, we realize that this set is by no means exhaustive. Different sets of metrics may be applicable depending on what the diagnosis results are supporting (abort decisions, ground support, fault-adaptive control, etc.). In addition there might be a set of weights associated with the metrics depending on their impor-

Name	LYDIA			NGDE			RODON		
	$\bar{M}_{\text{utl}}$	$\bar{M}_{\text{cpu}}$	$\bar{M}_{\text{mem}}$	$\bar{M}_{\text{utl}}$	$\bar{M}_{\text{cpu}}$	$\bar{M}_{\text{mem}}$	$\bar{M}_{\text{utl}}$	$\bar{M}_{\text{cpu}}$	$\bar{M}_{\text{mem}}$
74182	0.365	62	17	0.466	230	10 716	0.262	1 293	18 205
74L85	0.455	53	18	0.575	341	11 838	0.372	5 233	22 533
74283	0.419	57	17	0.479	206	10 654	0.353	4 863	20 714
74181	0.374	73	21	0.486	213	10 879	0.405	14 222	26 962
c432	0.529	91	24	0.664	319	12 058	0.492	19 129	36 772
c499	0.29	80	33	0.414	1 719	17 063	0.258	20 649	36 436
c880	0.262	1 842	37	0.296	1 516	21 437	0.275	18 404	34 843
c1355	0.335	387	34	0.37	4 734	23 967	0.373	22 133	33 653
c1908	0.208	745	29	0.232	8 994	33 995	0.19	24 361	36 102
c2670	0.603	327	119	0.921	571	14 828	0.886	17 178	34 069
c3540	0.355	833	33	0.374	9 223	31 954	0.307	49 397	48 162
c5315	0.243	811	94	0.531	6 477	22 406	0.238	87 720	50 526
c6288	0.316	2 162	32	0.32	11 784	65 086	0.316	89 130	51 268
c7552	0.3	2 001	97	0.436	8 638	39 592	0.364	172 558	65 846
Averaged	0.361	680	43	0.469	3 926	23 320	0.364	39 019	36 864

Table 12: Synthetic systems metrics results

	LYDIA			NGDE			RODON		
	$\bar{M}_{\text{sru}}$	$\bar{M}_{\text{dru}}$	$\bar{M}_{\text{err}}$	$\bar{M}_{\text{sru}}$	$\bar{M}_{\text{dru}}$	$\bar{M}_{\text{err}}$	$\bar{M}_{\text{sru}}$	$\bar{M}_{\text{dru}}$	$\bar{M}_{\text{err}}$
74182	0.381	0.984	69	0.574	0.892	78	0.262	1	80
74L85	0.458	0.996	30	0.617	0.958	39	0.46	0.913	78
74283	0.437	0.982	46	0.523	0.957	51	0.423	0.93	82
74181	0.378	0.995	48	0.517	0.969	55	0.456	0.949	87
c432	0.53	0.999	29	0.671	0.993	35	0.505	0.987	64
c499	0.293	0.997	71	0.428	0.986	78	0.268	0.99	107
c880	0.263	0.999	89	0.306	0.99	127	0.281	0.994	113
c1355	0.336	0.999	73	0.375	0.995	94	0.375	0.999	71
c1908	0.208	0.999	69	0.239	0.993	113	0.191	1	70
c2670	0.603	1	24	0.921	1	6	0.886	1	10
c3540	0.355	1	58	0.376	0.999	88	0.308	0.999	82
c5315	0.243	1	73	0.532	0.999	58	0.239	0.999	114
c6288	0.317	1	16	0.32	1	15	0.317	0.999	18
c7552	0.3	1	60	0.437	0.999	69	0.364	0.999	70
Averaged	0.364	0.996	54.02	0.488	0.981	64.75	0.381	0.983	74.71

Table 13: Synthetic systems secondary metrics results

tance (for abort decisions the fault detection time is of utmost importance). We expect to add more metrics to the list in the future (with support tools to compute those metrics). In addition since we were dealing with abrupt, persistent, and discrete faults, metrics associated with incipient, intermittent, and/or continuous faults were not considered.

Finally, the metrics listed in this paper do not capture the amount of effort necessary to build models of sufficient fidelity for the diagnosis task at hand. Furthermore, we have not investigated the ease or difficulty of updating models with new or changed system information. The art of building models is an important practical consideration which is not addressed in the current work.

In future work, we would like to determine a set of application-specific use cases (maintenance, autonomous operation, abort decision etc.) that the DA is supporting and select metrics that would be relevant to that use case.

5.4 Empirical Evaluation

Some practical issues arose in the execution of experiments. Much effort was put into ensuring stable, uniform conditions on the host machines; however, during the implementation, it was necessary to take measures that may have caused slight variability. One example was the manual examination of ongoing experiment results for quality assurance. Future releases of the DXF can address this by being more robust to unexpected DA behavior, and sending notifications in the event of such. Additionally, for Java DAs, significant dif-

Name	LYDIA SAT	UNSAT	$\bar{M}_{\text{sat}}$	NGDE SAT	UNSAT	$\bar{M}_{\text{sat}}$	RODON SAT	UNSAT	$\bar{M}_{\text{sat}}$
74182	19	45	5.67	1240	0	28	0	0	0
74L85	1	27	1	178	0	20	13	7	13
74283	34	57	5.32	561	0	20	15	5	15
74181	12	43	2.4	691	0	20	4	16	4
c432	10	29	4.7	1109	0	20	5	15	5
c499	2	118	1	707	0	20	2	12	2
c880	27	86	1.74	12663	0	20	15	0	15
c1355	36	162	4.1	3246	0	20	8	3	8
c1908	13	35	1	3593	4	7	0	2	0
c2670	7	30	7	25	0	19	17	2	17
c3540	38	77	1.86	231	10	10	1	17	1
c5315	0	55	0	1665	0	20	0	13	0
c6288	8	30	0.27	126	0	2	2	2	2
c7552	7	53	0.64	1493	3	17	1	17	1
Averaged	15.29	60.50	2.62	1966.29	1.21	17.36	5.93	7.93	5.93

Table 14: Synthetic systems satisfiability results

ferences were evident in the peak memory usage metric when run on Linux versus Windows. The problem was mostly bypassed by running all but one Java DA on Linux.

6 CONCLUSION

We presented a framework for evaluating and comparing DAs under identical conditions. The framework is general enough to be applied to any system and any kind of DA. The run-time architecture was designed to be as platform independent as possible. We defined a set of metrics that might be of interest when designing a diagnostic algorithm and the framework includes tools to compute the metrics by comparing actual scenarios and diagnostic results.

Using the framework, we have experimented with 13 diagnostic algorithms on 16 systems of various size and synthetic/real-world origin. We have, both manually and programatically, created 1651 observation scenarios of various complexity. We have designed 10 metrics for measuring diagnostic performance. This has resulted in the execution of 6484 scenarios with a total duration of more than 169.7 hours and the computation of 84292 metrics.

We presented the results from our effort to evaluate the performance of a set of diagnostic algorithms on the ADAPT electrical power system testbed, and a set of synthetic circuits. We learned valuable lessons in trying to complete this effort. One major take-away is that there is still a lot of work and discussion needed to determine a common comparison and evaluation framework for the diagnosis community. The other key observation is that no DA was able to be best in a majority of the metrics. This clearly indicates that the selection of DAs would necessarily involve a trade-off analysis between various performance metrics.

The framework presented is by no means a finished product and we expect it to evolve over the years. In the paper, we have identified some of the limitations and expected scope for future expansion. Our sincere hope is that the framework is adopted by growing number of people and applied to a wide variety of physical systems including diagnosis algorithms from several different research communities. The long-term goal is to create a database of performance evaluation results which will allow system designers to choose the appropriate DA for their system given the constraints and metrics in their application.

ACKNOWLEDGEMENTS

We extend our gratitude to Gautam Biswas (Vanderbilt University), Kai Goebel (University Space Research Association), Ole Mengshoel (Carnegie Mellon University), Gregory Provan (University College Cork), Peter Struss (Technical University Munich), Serdar Uckun (PARC), and many others for valuable discussions in establishing the evaluation framework. In addition, we extend our gratitude to David Hall (Stinger Ghaffarian Technologies), David Jensen (Oregon State University), David Nishikawa (NASA), Brian Ricks (University of Texas at Dallas), Adam Sweet (NASA), Michel Wilson (Delft University of Technology), Stephanie Wright (Vanderbilt University), and many others for supporting the work reported here.

This research was supported in part by the National Aeronautics and Space Administration (NASA) Aeronautics Research Mission Directorate (ARMD) Aviation Safety Program (AvSP) Integrated Vehicle Management (IVHM) Project. Additionally, this material is based upon the work supported by NASA under award NNA08CG83C.

This research was supported by PROGRESS, the embedded systems research program of the Dutch organisation for Scientific Research NWO, the Dutch Ministry of Economic Affairs and the Technology Foundation STW under award DES.07015.

NOMENCLATURE

IN	inputs
OUT	outputs
COMPS	components
V	variables
C	clauses
t_d	first detection
t_i	last isolation
C_s	startup CPU cycles
C	CPU cycles per step
M	memory in use
$\omega^\star$	injected fault
$t_i^\star$	injection of fault i
Ω	candidate diagnoses
$\Omega^\top$	satisfiable candidate diagnoses
W	candidate weights
f	number of all components
n	number of false negatives
N	number of healthy components
$\bar{n}$	number of false positives
$\bar{N}$	number of faulty components
m_{ia}	candidate isolation accuracy
m_{sru}	candidate system repair utility
m_{dru}	candidate diagnosis repair utility
m_{utl}	candidate utility
M_{fd}	scenario fault detection time
M_{fn}	scenario false negative
M_{fp}	scenario false positive
M_{da}	scenario detection accuracy
M_{fi}	scenario fault isolation time
M_{ia}	scenario isolation accuracy
M_{err}	scenario classification errors
M_{utl}	scenario utility
M_{sru}	scenario system repair utility
M_{dru}	scenario diagnosis repair utility
M_{sat}	scenario consistency
M_{cpu}	scenario CPU load
M_{mem}	scenario memory load
$\bar{M}_{fd}$	system fault detection time
$\bar{M}_{fn}$	system false negative
$\bar{M}_{fp}$	system false positive
$\bar{M}_{da}$	system detection accuracy
$\bar{M}_{fi}$	system fault isolation time
$\bar{M}_{err}$	system classification errors
$\bar{M}_{utl}$	system utility
$\bar{M}_{sru}$	system system repair utility
$\bar{M}_{dru}$	system diagnosis repair utility
$\bar{M}_{sat}$	system consistency
$\bar{M}_{cpu}$	system CPU load
$\bar{M}_{mem}$	system memory load

REFERENCES

Bartyś, M., Patton, R., Syfert, M., de las Heras, S., & Quevedo, J. (2006). Introduction to the DAMADICS actuator FDI benchmark study. *Control Engineering Practice, 14*, 577–596.

Basseville, M., & Nikiforov, I. V. (1993). *Detection of Abrupt Changes: Theory and Application*. Prentice-Hall.

Brglez, F., & Fujiwara, H. (1985). A neutral netlist of 10 combinational benchmark circuits and a target translator in Fortran. In *Proc. IS-CAS'85*, pp. 695–698.

Committee E-32, S. A. P. S. H. M. (2008). Health and usage monitoring metrics, monitoring the monitor. Tech. rep. ARP5783.

Davis, M., Logemann, G., & Loveland, D. (1962). A machine program for theorem-proving. *Communications of the ACM, 5*(7), 394–397.

de Freitas, N. (2002). Rao-blackwellised particle filtering for fault diagnosis. In *Proc. AERO-CONF'02*, Vol. 4, pp. 1767–1772.

de Kleer, J. (2009). Minimum cardinality candidate generation. In *Proc. DX'09*, pp. 397–402.

de Kleer, J., Mackworth, A., & Reiter, R. (1992). Characterizing diagnoses and systems. *Artificial Intelligence, 56*(2-3), 197–222.

de Kleer, J., & Williams, B. (1987). Diagnosing multiple faults. *Artificial Intelligence, 32*(1), 97–130.

DePold, H. R., Rajamani, R., Morrison, W. H., & Pattipati, K. R. (2006). A unified metric for fault detection and isolation in engines. In *Proc. TURBO'06*.

DePold, H. R., Siegel, J., & Hull, J. (2004). Metrics for evaluating the accuracy of diagnostic fault detection systems. In *Proc. TURBO'04*.

Feldman, A., Provan, G., & van Gemund, A. (2007). Interchange formats and automated benchmark model generators for model-based diagnostic inference. In *Proc. DX'07*, pp. 91–98.

Feldman, A., Provan, G., & van Gemund, A. (2008a). Computing minimal diagnoses by greedy stochastic search. In *Proc. AAAI'08*, pp. 911–918.

Feldman, A., Provan, G., & van Gemund, A. (2008b). Computing observation vectors for max-fault min-cardinality diagnoses. In *Proc. AAAI'08*, pp. 911–918.

Feldman, A., Provan, G., & van Gemund, A. (2009). The Lydia approach to combinational model-based diagnosis. In *Proc. DX'09*, pp. 403–408.

Forbus, K., & de Kleer, J. (1993). *Building Problem Solvers*. MIT Press.

Gertler, J. J. (1998). *Fault Detection and Diagnosis in Engineering Systems*. Marcel Dekker.

Grastien, A., & Kan-John, P. (2009). Wizards of Oz – description of the 2009 DXC entry. In *Proc. DX'09*, pp. 409–413.

Heller, U., & Struss, P. (2001). G$^+$DE - the generalized diagnosis engine. In *Proc. DX'01*, pp. 79–86.

Hoyle, C., Mehr, A. F., Tumer, I. Y., & Chen, W. (2007). Cost-benefit quantification of ISHM in aerospace systems. In *Proc. IDETC/CIE'07*.

Iverson, D. L. (2004). Inductive system health monitoring. In *Proc. ICAI'04*.

Izadi-Zamanabadi, R., & Blanke, M. (1999). A ship propulsion system as a benchmark for fault-tolerant control. *Control Engineering Practice*, *7*(2), 227–240.

Karin, L., Lunde, R., & Münker, B. (2006). Model-based failure analysis with RODON. In *Proc. ECAI'06*.

Kavčič, M., & Juričić, Đ. (1997). A prototyping tool for fault tree based process diagnosis. In *Proc. DX'97*, pp. 129–133.

Kurien, J., & Moreno, M. D. R. (2008). Costs and benefits of model-based diagnosis. In *Proc. AEROCONF'08*.

Kurtoglu, T., Mengshoel, O., & Poll, S. (2008). A framework for systematic benchmarking of monitoring and diagnostic systems. In *Proc. PHM'08*.

Kurtoglu, T., Narasimhan, S., Poll, S., Garcia, D., Kuhn, L., de Kleer, J., van Gemund, A., & Feldman, A. (2009). First international diagnosis competition - DXC'09. In *Proc. DX'09*, pp. 383–396.

Larsson, J. E. (1996). Diagnosis based on explicit means-end models. *Artificial Intelligence*, *80*(1).

Lerner, U., Parr, R., Koleer, D., & Biswas, G. (2000). Bayesian fault detection and diagnosis in dynamic systems. In *Proc. AAAI'00*, pp. 531–537.

Mengshoel, O. (2007). Designing resource-bounded reasoners using Bayesian networks: System health monitoring and diagnosis. In *Proc. DX'07*, pp. 330–337.

Metz, C. E. (1978). Basic principles of ROC analysis. *Nuclear Medicine*, *8*(4), 283–298.

Narasimhan, S., & Brownston, L. (2007). HyDE - a general framework for stochastic and hybrid modelbased diagnosis. In *Proc. DX'07*, pp. 162–169.

Orsagh, R. F., Roemer, M. J., Savage, C. J., & Lebold, M. (2002). Development of performance and effectiveness metrics for gas turbine diagnostic technologies. In *Proc. AEROCONF'02*, Vol. 6, pp. 2825–2834.

Poll, S., Patterson-Hine, A., Camisa, J., Garcia, D., Hall, D., Lee, C., Mengshoel, O., Neukom, C., Nishikawa, D., Ossenfort, J., Sweet, A., Yentus, S., Roychoudhury, I., Daigle, M., Biswas, G., & Koutsoukos, X. (2007). Advanced diagnostics and prognostics testbed. In *Proc. DX'07*, pp. 178–185.

Reiter, R. (1987). A theory of diagnosis from first principles. *Artificial Intelligence*, *32*(1), 57–95.

Roemer, M., Dzakowic, J., Orsagh, R. F., Byington, C. S., & Vachtsevanos, G. (2005). Validation and verification of prognostic health management technologies. In *Proc. AEROCONF'05*, pp. 3941–3947.

Roychoudhury, I., Biswas, G., & Koutsoukos, X. (2009). Designing distributed diagnosers for complex continuous systems. *IEEE Trans. on Automation Science and Engineering*, *6*(2), 277–290.

Russell, S., & Norvig, P. (2003). *Artificial Intelligence: A Modern Approach*. Pearson Education.

Schuster, E. F., & Sype, W. R. (1987). On the negative hypergeometric distribution. *International Journal of Mathematical Education in Science and Technology*, *18*(3), 453–459.

Siddiqi, S., & Huang, J. (2007). Hierarchical diagnosis of multiple faults. In *Proc. IJCAI'07*, pp. 581–586.

Simon, L., Bird, J., Davison, C., Volponi, A., & Iverson, R. E. (2008). Benchmarking gas path diagnostic methods: A public approach. In *Proc. ASME Turbo Expo 2008*.

Sorsa, T., & Koivo, H. (1998). Application of artificial neural networks in process fault diagnosis. *Automatica*, *29*(4), 843–849.

Williams, Z. (2006). Benefits of IVHM: An analytical approach. In *Proc. AEROCONF'06*, pp. 1–9.

Zymnis, A., Boyd, S., & Gorinevsky, D. (2009). Relaxed maximum a posteriori fault identification. *Signal Processing*, *89*(6), 989–999.

A DERIVATIONS OF METRICS

This appendix provides detailed derivation of the formulae for the technical accuracy metrics. In this appendix we use notation of Sec. 3.3 (in particular, recall Fig. 4 and Table 3).

A.1 Classification Errors and Isolation Accuracy

Recall the definition of M_{err} and M_{ia}:

$$M_{\text{err}} = \sum_{\omega \in \Omega} W(\omega)(|\omega \ominus \omega^\star|) \qquad (20)$$

$$M_{\text{ia}} = \sum_{\omega \in \Omega} W(\omega)(f - |\omega \ominus \omega^\star|) \qquad (21)$$

One can see that M_{ia} and M_{err} are duals, i.e.:

$$\frac{M_{\text{ia}}}{f} + \frac{M_{\text{err}}}{f} = 1 \qquad (22)$$

Consider the isolation accuracy (m_{ia}) of a single diagnostic candidate $\omega \in \Omega$:

$$m_{\text{ia}} = f - |\omega \ominus \omega^\star| \qquad (23)$$

Eq. 23 defines a plane in the $(n, \bar{n}, m_{\text{ia}})$-space (see Fig 13).

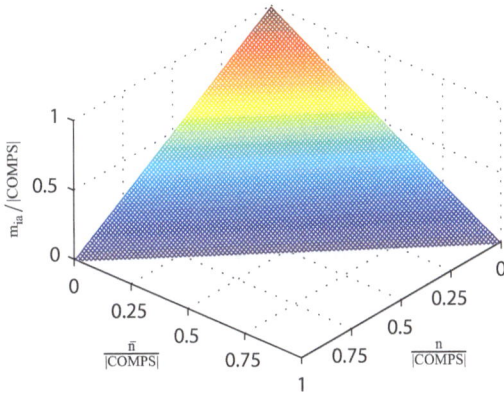

Figure 13: m_{ia} as a function of n and $\bar{n}$

m_{ia} "penalizes" a DA for each misclassified component. As is visible from Fig. 13, the penalty is applied linearly.

The isolation accuracy metric M_{ia} originates in the automotive industry (Committee E-32, 2008). The Aerospace Recommended Practice (ARP) computes the closely related probability of correct classification in the following way. For each component we compute the square confusion matrix. The probability of correct classification is the sum of the main diagonal divided by the total number of classifications (see the referenced ARP (Committee E-32, 2008) for details and examples).

It can be shown that the probability of correct classification, as defined in the above ARP, is equivalent to M_{ia}, if both fault and nominal component modes are used for the computation of the confusion matrices. The probability of correct classification is conditioned on the fault probability while the probability measured by M_{ia} is

not. The latter is purely a metric design consideration. The fact that we use nominal modes for computing M_{ia} leads to higher correlation of M_{ia} with the detection accuracy metrics defined later in this section.

If more than one predicted mode vector is reported by a DA, (meaning that the diagnostic output consists of a set of candidate diagnoses), then the isolation accuracy and the classification errors are calculated for each predicted component mode vector and weighted by the candidate probabilities reported by the DA as it is seen in Eq. (20) and Eq. (14). M_{ia} and M_{err} are very useful for single diagnoses but with multiple candidates they are less intuitive. The metric that follows is loosely based on the concept of "repair effort" and partly remedies this problem.

A.2 Utilities

In what follows we show the derivations of the three utility metrics (system repair utility M_{sru}, diagnosis repair utility M_{dru}, and utility M_{utl}).

A.2.1 System Repair Utility

Consider an injected fault $\omega^\star$ ($\omega^\star$ is a set of faulty components) and a diagnostic candidate ω (the set of components the DA considers faulty). The number of truly faulty components that are improperly diagnosed by the diagnostic algorithm as healthy (false negatives) is $n = |\omega^\star \setminus \omega|$ (see Fig. 4). In general a diagnostician has to perform extra work to verify a diagnostic candidate ω, which must be reflected in the system repair utility. We assume that he or she has access to a test oracle that reports if a component c is healthy or faulty.

We first determine what the expected number of tests a diagnostician has to perform to test all components in $\omega^\star \setminus \omega$ (the false negatives) if the diagnostician chooses untested components at random with uniform probability. In the worst case, the diagnostician has to test all the remaining $\text{COMPS} \setminus \omega$ components (the diagnostic algorithm has already determined the state of all components in ω). Consider the average situation. We denote $N = |\text{COMPS} \setminus \omega|$. N is the size of the "population" of components to be tested.

The probability of observing $s - 1$ successes (faulty components) in $k + s - 1$ trials (i.e., k oracle tests) is given by the direct application of the hypergeometric distribution:

$$p(k, s - 1) = \frac{\binom{n}{s-1}\binom{N-n}{k}}{\binom{N}{k+s-1}} \qquad (24)$$

The probability $p(k, s)$ of then observing a faulty component in the next oracle test is simply the number of remaining false negatives $n - (s - 1)$ divided by the size of the remaining population $(N - (s + k - 1))$:

$$p(k, s) = \frac{n - s + 1}{N - k - s + 1} \qquad (25)$$

and the probability of having exactly k oracle faults up to the sth test, is then the product of these two probabilities:

$$p'(k,s,n,N) = \frac{\binom{n}{s-1}\binom{N-n}{k}(n-s+1)}{\binom{N}{k+s-1}(N-k-s+1)} \quad (26)$$

The formula above is the probability mass of the inverse hypergeometric distribution that, in our case, yields the probabilities for testing k healthy components before we find s faulty components out of the population (no repetitions). The expected value $E'[k]$ of $p'(k,s,n,N)$ (from the definition of a first central moment of a random variable) is:

$$E'[k] = \sum_{x=0}^{n} x p'(x,s,n,N) \quad (27)$$

Replacing $p'(k,s,n,N)$ in (27) and simplifying gives us the mean of the inverse hypergeometric distribution[2]:

$$E'[k] = \frac{s(N-n)}{n+1} \quad (28)$$

As we are interested in finding $s = n$ faulty components, the expected value $E'(n,N)$ becomes:

$$E'[k] = \frac{n(N-n)}{n+1} \quad (29)$$

The expected number of tests $E[t]$ (as opposed to the expected number of faulty components $E'[k]$) then becomes:

$$E[t] = \frac{n(N-n)}{n+1} + n = \frac{n(N+1)}{n+1} \quad (30)$$

The expected number of tests $E[t]$ is then normalized by the number of components f and flipped alongside the y axis to give the system repair utility:

$$m_{\text{sru}} = 1 - \frac{n(N+1)}{f(n+1)} \quad (31)$$

Plotting the system repair utility m_{sru} against a variable number of false negatives is shown in Fig. 14. One can see that unlike m_{err} which changes linearly, m_{sru} "penalizes" improperly diagnosed components exponentially.

The system repair utility for a set of diagnoses is defined as:

$$M_{\text{sru}} = \sum_{\omega \in \Omega} W(\omega) m_{\text{sru}}(\omega^\star, \omega) \quad (32)$$

where $W(\omega)$ is the weight of a diagnosis ω such that:

$$\sum_{\omega \in \Omega} W(\omega) = 1 \quad (33)$$

All weights $W(\omega)$, $\omega \in \Omega$, are computed by the diagnostic algorithm.

---

[2]For a detailed derivation of the negative hypergeometric mean, see (Schuster & Sype, 1987).

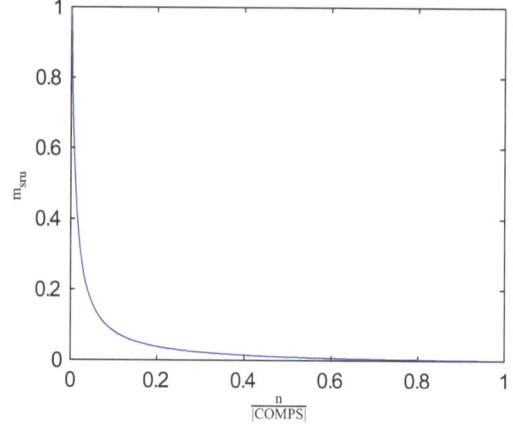

Figure 14: m_{sru} as a function of n

A.3 Diagnosis Repair Utility

Using $E[t]$ in a metric is not enough as it only captures the effort to "eliminate" (test) all false negatives. The size of the set of false positives is $\bar{n} = |\omega \setminus \omega^\star|$ (see Fig. 4). To find all false positives, the diagnostician has to test in the worst case all components in ω. Hence, the general population is $\bar{N} = |\omega|$. Repeating the argument for $E[t]$ we determine the expected number of tests for testing all false positives $E[\bar{t}]$:

$$E[\bar{t}] = \frac{\bar{n}(\bar{N}+1)}{\bar{n}+1} \quad (34)$$

Similarly, the diagnostic repair utility m_{dru} is the normalized $E[\bar{t}]$:

$$m_{\text{dru}} = 1 - \frac{\bar{n}(\bar{N}+1)}{f(\bar{n}+1)} \quad (35)$$

The diagnosis repair utility for a set of diagnoses is defined as:

$$M_{\text{dru}} = \sum_{\omega \in \Omega} W(\omega) m_{\text{dru}}(\omega^\star, \omega) \quad (36)$$

A.4 Utility

The utility metric (per candidate) is a combination of m_{sru} and m_{dru}:

$$m_{\text{utl}} = 1 - \frac{E[t] + E[\bar{t}]}{f} = \quad (37)$$

$$= 1 - \frac{n(N+1)}{f(n+1)} - \frac{\bar{n}(\bar{N}+1)}{f(\bar{n}+1)} \quad (38)$$

The utility metric (per scenario) is

$$M_{\text{utl}} = \sum_{\omega \in \Omega} W(\omega) m_{\text{utl}}(\omega^\star, \omega) \quad (39)$$

Figure 15 plots m_{utl} for varying numbers of false negatives and false positives in a (symmetric) case where the cardinality of the injected fault is half the number of components. Normally, the number of injected faulty components $|\omega^\star|$ is small compared to the total number of components f, which leads to an asymmetric m_{utl} plot. In such cases, $\bar{N} \ll N$, hence the role of the false positives is small. In Fig. 15, there is a global optimum $m_{utl} = 1$ for $n = 0$ and $\bar{n} = 0$, i.e., all components in ω are classified correctly.

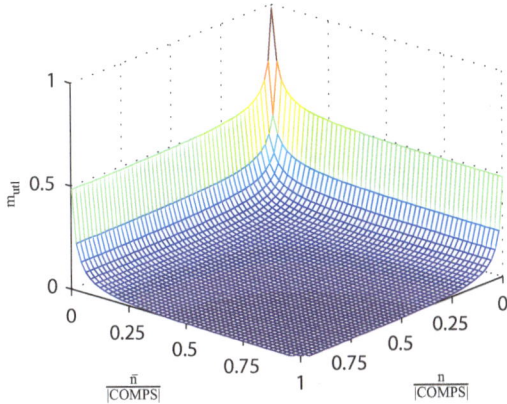

Figure 15: m_{utl} as a function of n and $\bar{n}$

B SYSTEM DESCRIPTION FORMAT

Consider c17, the smallest ISCAS85 circuit (too simple to include in our benchmark). An example system description starts by defining a number of components in the following manner (we have truncated the XML code):

```
<?xml version="1.0" encoding="UTF-8"?>

<systemCatalog ...>
  <systemInstances>
    <systemInstance id="c17" system="c17" />
  </systemInstances>
  <systems>
    <system>
      <systemName>c17</systemName>
      <description>
        The c17 ISCAS85 combina-
tional circuit.
      </description>
      <components>
        <component>
          <name>i1</name>
          <componentType>port</componentType>
        </component>
        <component>
          <name>gate11</name>
          <componentType>nand2</componentType>
        </component>
        <component>
          <name>gate11.o</name>
          <componentType>wire</componentType>
        </component>
          ⋮
```

Part of the topology of c17 is described in the XML excerpt below:

```
<connections>
  <connection>
    <c1>gate10.o</c1><c2>z1</c2>
  </connection>
  <connection>
    <c1>gate10.i1</c1><c2>i1</c2>
  </connection>
  <connection>
    <c1>gate10.i2</c1><c2>i3</c2>
  </connection>
    ⋮
<connections>
```

The component type specifying a circuit breaker and shown next is part of ADAPT-Lite and ADAPT (this component type is referenced, for example, by a component with unique identifier CB180):

```
<componentType xsi:type="circuitBreaker">
  <name>CircuitBreaker4Amp</name>
  <description>
    4 Amp CircuitBreaker
  </description>
  <modesRef>CircuitBreaker</modesRef>
  <rating>4</rating>
</componentType>
```

Another example of a component type is the AC voltage sensor shown below.

```
<componentType xsi:type="sensor">
  <name>ACVoltageSensor</name>
  <description>
    AC voltage sensor.
  </description>
  <modesRef>ScalarSensor</modesRef>
  <sensorValue xsi:type="numberValue">
    <dataType>double</dataType>
    <rangeMin>0</rangeMin>
    <rangeMax>150</rangeMax>
  </sensorValue>
  <engUnits>VAC</engUnits>
</componentType>
```

Below is shown a nand-gate, part of a digital circuit.

```
<componentType>
  <name>nand2</name>
  <description>
    A 2-input logic NAND gate.
  </description>
  <modesRef>gate</modesRef>
</componentType>
```

Finally, we have the modes of a circuit-breaker.

```
<modeGroup>
  <name>CircuitBreaker</name>
  <mode xsi:type="mode">
    <name>Nominal</name>
    <description>
      Transmits current and voltage ...
    </description>
  </mode>
  <mode xsi:type="mode">
    <name>Tripped</name>
    <description>
```

```
    Breaks the circuit and must be ...
  </description>
</mode>
<mode xsi:type="faultMode">
  <name>FailedOpen</name>
  <description>
    Trips even though current is ...
  </description>
  <faultSource>Hardware</faultSource>
  <parameters/>
</mode>
</modeGroup>
```

C MESSAGE FORMATS

Though there are additional message types, the most important messages for the purpose of benchmarking are the sensor data message, command message, and diagnosis message, described below.

C.1 Sensor/Command Data

Sensor data are defined broadly as a map of sensor IDs to sensor values (observations). Sensor values can be of any type; currently the framework allows for integer, real, boolean, and string values. The type of each observation is indicated by the system's XML catalog.

SensorMessage
+timestamp
+sensorValues: Map<sensorIds→sensorValues>

CommandMessage
+timestamp
+commandID: string
+command: commandValue

Table 15: Sensor and command message format

Commandable components contain an additional entry in the system catalog specifying a command ID and command value type (analogous to sensor value type). The command message represents the issuance of a command to the system. In the ADAPT system, for example, the message (EY144_CL, true) signifies that relay EY144 is being commanded to close. "EY144_CL" is the command ID, and "true" is the command value (in this case, a Boolean).

C.2 Diagnosis Result Format

The DA's output (i.e., estimate of the physical status of the system) is standardized to facilitate the generation of common data sets and the calculation of the benchmarking metrics, which are introduced in Sec. 3.3. The resulting diagnosis message is summarized in Table 16 and contains:

timestamp: a value indicating when the diagnosis has been issued by the algorithm.

candidateSet: a candidate fault set is a list of candidates an algorithm reports as a diagnosis. A candidate fault set may include a single candidate with a single or multiple faults; or multiple candidates each with a single or multiple faults. It is assumed that only one candidate in a candidate fault set can represent the system at any given time.

detectionSignal: a Boolean value as to whether the diagnosis system has detected a fault.

isolationSignal: a Boolean value as to whether the diagnosis system has isolated a candidate or a set of candidates.

DiagnosisMessage
+timestamp
+candidateSet: Set <Candidate>
+detectionSignal: Boolean
+isolationSignal: Boolean
+notes: string

Candidate
+faults: Map<componentIds→componentState>
+weight: double

Table 16: Diagnosis message format

In addition, each candidate in the candidate set has an associated weight. Candidate weights are normalized by the framework such that their sum for any given diagnosis is 1.

A Survey of Health Management User Objectives in Aerospace Systems Related to Diagnostic and Prognostic Metrics

Kevin R. Wheeler [1], Tolga Kurtoglu [2], and Scott D. Poll [1]

[1] NASA Ames Research Center, Moffett Field, CA, 94035, USA
kevin.r.wheeler@nasa.gov
scott.d.poll@nasa.gov

[2] Palo Alto Research Center, Palo Alto, CA, 94304, USA
kurtoglu@parc.com

ABSTRACT

One of the most prominent technical challenges to effective deployment of health management systems is the vast difference in user objectives with respect to engineering development. In this paper, a detailed survey on the objectives of different users of health management systems is presented. These user objectives are then mapped to the metrics typically encountered in the development and testing of two main systems health management functions: diagnosis and prognosis. Using this mapping, the gaps between user goals and the metrics associated with diagnostics and prognostics are identified and presented with a collection of lessons learned from previous studies that include both industrial and military aerospace applications.

1. INTRODUCTION

One of the possible reasons for slow adoption of integrated health management systems is the vast difference in user objectives with respect to engineering development. In this paper, we present a survey of the objectives of different users of integrated health management systems, how they each would measure success of such systems (metrics), and how these objectives and metrics relate to engineering efforts developing prognostic and diagnostic algorithms and systems. These user objectives and

This is an open-access article distributed under the terms of the Creative Commons Attribution 3.0 United States License, which permits unrestricted use, distribution, and reproduction in any medium, provided the original author and source are credited.

Submitted 3/2010; Published 10/2010

associated metrics are identified across operational, regulatory and engineering domains for both industrial and military aerospace applications.

This survey was sponsored by NASA's Aviation Safety Program, Integrated Vehicle Health Management (IVHM) Project to aid in identifying critical gaps within their existing research portfolio that are not currently being addressed by the broader research community.

2. ORGANIZATION

The rest of the paper is organized as follows: Section III gives background on the application of health management in the aviation domain; Section IV discusses the motivation and competing challenges for health management; Section V presents objectives and metrics for different health management users; Section VI describes metrics associated with development and operation of diagnostic and prognostic systems; finally, sections VII and VIII provide discussion and summary, respectively, of the topics in this paper.

3. BACKGROUND

The first generation aircraft health management system (as exemplified in B727, DC-9/MD-80, B737 classic) consisted of "push-to-test" functionality of mechanical and analog systems in which a button was pressed to test internal circuitry and simple status lights would illuminate the go/no-go results for the device under test. The second generation (B757/767, B737NG, MD-90, A320) saw the use of black-box digital systems to carry out the health management functions previously performed by mechanical and analog systems. The third generation (MD-11, B747-400) saw the introduction of systems implementing the ARINC Standard 604, "Guidance for Design and Use of

Built-In Test Equipment." Early third generation systems allowed centralized access to the federated avionics BIT results but required manual consolidation of Line Replaceable Unit (LRU) fault indications. Later third generation systems used Central Maintenance Computers to aggregate all fault indications and perform root cause analysis via complex logic-based equations. The ability to downlink fault results to ground stations while *en route* was also added. Lessons learned were incorporated into updated standards, including ARINC 624, "Design Guidance for Onboard Maintenance System." The fourth generation implements improved health management functionality through the use of modular avionics. In contrast to having specific avionics functions associated with a LRU, multiple avionics functions are associated with Line Replaceable Modules. The health management system employed on the Boeing 777 represents the fourth generation in the evolution of vehicle health management (Honeywell, 2007). The Boeing 777 Airplane Information Management System integrates two key diagnostic subsystems: the Central Maintenance Computing Function, which diagnoses faults after they happen, and the Airplane Condition Monitoring Function, which collects data to allow prediction of future problems and thus enables condition-based maintenance. In contrast to the logic equation-based diagnostics in previous health management systems, the central maintenance control system in the Boeing 777 employs model-based reasoning in an attempt to overcome difficulties in developing and maintaining the health management functions. Subsequent developments have extended the scalability and extensibility of the modular avionics systems and the associated health management functionality. Despite the advances over the years, there are still difficulties in developing and implementing health management systems that meet user requirements (Scandura, 2005), although these difficulties may be more programmatic than technical.

MacConnell (2007) conducted an extensive working group study on the benefits of ISHM consisting of representatives from the Air Force Research Laboratory (AFRL), Boeing, General Electric, Honeywell, Lockheed Martin, Northrop Grumman, United Technologies, and others. New benefits were identified that may be perceived as more indirect. For example, automated monitoring could be relied on to dramatically reduce factors of safety for design and to enable revolutionary certification processes. The working group (MacConnell, 2007) ranked the relative importance of the functional areas in ISHM benefits. The top five were diagnostics, analysis, design, structure and propulsion. This is a mix of health management functions (diagnostics, analysis) with application areas (structure, propulsion). This highlights that the words used (ontology) amongst even a group of experts can cause opacity in health management discussions thus making it difficult to clearly outline the requirements driving the development and integration of fleet wide health management systems.

Ofstun (2002) has a succinct overview of developing IVHM for aerospace platforms, pointing out that traditional built-in-tests generally have not provided the accuracy or reliability needed to impact operational efficiency and maintenance. A goal of IVHM should be to both improve and extend traditional BIT approaches in subsystems such as avionics, electrical (including wiring), actuators, environmental control, propulsion, hydraulics, structures as well as overall system performance.

Ofsthun (2002) highlights IVHM lessons learned that are points similar to those that will be seen in this article relating user community goals to diagnostic and prognostic modeling metrics. Our article highlights:

IVHM performance measures need to be derived by an integrated product development team that accounts for all expected user groups.

• Cost/benefit analyses need to be conducted for each expected user group during requirements definition.

• A common health management infrastructure is needed to integrate across subsystems - including definition of subsystem responsibilities.

• Trade-space analyses need to be conducted between failure detection and false alarm rates – including crew enabled filtering.

• Verification and validation of IVHM system needs to include incremental validation by demonstrations as well as opportunistic monitoring.

Currently the best developing example of a highly integrated system for health management is the Joint Strike Fighter program (JSF) which mandates such a development (Hess et al., 2004). One of the greatest challenges in developing a health management system from the ground up has been in refining the user objectives and requirements to an adequate level that includes buy-in from the expected and varied user groups.

The following section outlines general (non-formatting) guidelines to follow. These guidelines are applicable to all authors and include information on the policies and practices relevant to the publication of your manuscript.

3. MOTIVATION

Wide-spread adoption of integrated health management has been slow due to competing factors that have to be satisfied within the HM user community. Two areas stand out in this regard: Aging and Expected Life and Cost vs. Benefit.

3.1 Aging and Expected Life

As the average age of air fleets begins to be higher than the original expected useful life, in order to preserve safety-of-flight, it becomes necessary to increase the periodicity and depth of inspections.

This results in an increase in maintenance costs as well as longer periods of downtime. One of the benefits of an ISHM system that includes structural health monitoring is that this inspection burden can be reduced by relying upon continuous monitoring (Albert et al., 2006). The USAF has deployed structural monitoring systems that allow for the required maintenance inspection interval to be tailored to each aircraft, which has resulted in reducing the inspection burden, costs and amount of downtime.

One might be tempted to suggest that if the average age of an air fleet (either military or commercial) is starting to exceed the expected life, then a possible strategy to reduce the average age would be to begin replacement of the oldest with new aircraft. Unfortunately, especially in the case of the U.S. DoD, with given budgets it would not be possible to decrease the average age enough to make a difference. This is also true in civilian fleets: "The statistics show that the number of aging aircraft (older than 15 years) has increased continuously. This number was around 4600 in 1997 for US and European built civil aircraft flown with more than 1900 aircraft older than 25 years. This number increased to 4730 (>15 years) and 2130 (>25 years) respectively in 1999" (Boller, 2001).

From an engineering perspective, the development of health management systems design to mitigate the greatest risks is dependent upon accurate data collection. The data needed for maturation analysis is usually difficult both to obtain (due to heterogeneous systems) as well as to collect: "- this makes access, retrieval, and integration of the requisite information a costly and often incomplete process at best" (Wilmering, 2003).

3.2 Cost vs. Benefit

Installation of integrated health management systems incur development, installation and life cycle costs. Some of the costs associated with a health management solution include maintenance of the health management system components (such as sensor replacement and software upgrades) as well as increases in system volume and mass requirements. These costs need to be countered with expected savings gains over the life of the aircraft through a rigorous cost benefit analysis (CBA). The slow acceptance of health management tools has been attributed to the incomplete total life cycle systems engineering management (Millar, 2007) which introduces an approach for proper system analysis methods. Often the optimization of objectives consists of conflicting goals such as minimizing purchase cost and maximizing availability (Yukish et al., 2001). Calculating costs such as operating costs consists of complex parameters such as average downtime for unplanned repairs.

In spite of these challenges, different methods have been developed to analyze cost-benefit tradeoffs for designing and implementing IVHM systems. For example, [20] discusses the benefits of IVHM to five different categories of operators: the Original Equipment Manufacturers (OEMs), the mission operators, command/control elements, fleet management, and maintenance operators. These five categories may overlap in organizational structure and personnel, but they have clearly identifiable

processes and performance that can be measured. Another example of an approach to conducting a CBA for IVHM appears in (Ashby and Byer, 2002). Their methodology utilizes pre-existing reliability and logistics source information from a Failure Modes and Effects Criticality Analysis (FMECA), line maintenance activities and legacy field event rates. IVHM will have the greatest benefit when it is applied to those areas that are historically the least reliable, have failure modes that can greatly impact mission success, have sub-systems that are the most difficult to diagnose or for which replacements parts cannot be obtained in a timely-manner (Banks et al., 2005).

The impacts of diagnostic capability on unscheduled maintenance include (Ashby and Byer, 2002):

- reduction of cannot duplicate rates

- reduction of labor mean-time-to-detect (MTTD)

- reduction of line replaceable unit (LRU) repair costs

- reduction of repair times (increase availability)

The benefits impacting scheduled maintenance include:

- reduction of labor

- reduction of maintenance induced failures

- elimination of scheduled maintenance

Prognostic capabilities impacting operations include:

- reduction in number of engine in-flight shutdowns, mission aborts, lost sorties

- reduction of secondary damage

- ability to reconfigure and re-plan for optimal usage of the remaining useful life (RUL) of failing components

- maximized usage of the component life while ensuring mission safety

One example of cost-benefit quantification of ISHM in aerospace systems appears in (Hoyle et al., 2007). Their methodology analyzes the trade-offs between system availability, cost of detection, and cost of risk. In this optimization formulation, cost of detection includes the cost of periodic inspection/maintenance and the cost of ISHM; cost of risk quantifies risk in financial terms as a function of the consequential cost of a fault and the probabilities of occurrence and detection. Increasing the ISHM footprint will generally lower cost of risk while raising cost of detection, while availability will increase or decrease based upon the balance of the reliability and detection capability of the sensors added, versus their ability to reduce total maintenance time.

The business case for ISHM generated by an ISHM working group composed mostly of industry (MacConnell, 2007) resulted in the following rankings of benefits:

1. Maintenance time savings

2. False alarm avoidance - reduce can not duplicate (CND) and retest okay (RTOK)

3. Availability Improvement - increase MTBMA - mean time between maintenance actions

4. Spares and supply savings

5. Recurring cost savings.

In the past, there have been many anecdotal accounts of the benefits of ISHM. Now some systems, such as the condition based maintenance helicopter programs are starting to produce real results. For example in (Nenniger, 2007) implementing health management in the UH-60 has resulted in an increase in fully mission capable status from 65% to 87% resulting in an increase in total flight hours from 10,331 to 21,819.

There are uncertainties inherent to new Prognostics and Health Management (PHM) systems such as the fact that not all faults will be diagnosed correctly (PHM Effectiveness). Two factors that may detract from the benefits of prognostics (Hecht, 2006):

• Prognostics may cause some sub-systems to be replaced much earlier than their eventual failure thus reducing their useful life. This will require engineering resources to analyze replaced units in order to optimize replacement thresholds.

• False prognostic replacement indicators may cause replaceable units to be replaced that are not in any danger of failing. This will require further engineering resources to mitigate these false alarms.

The perceived and real difficulties of retrofitting legacy aviation systems with effective health management and the challenges of unambiguously quantifying the benefit in new systems has hampered more wide-spread adoption of integrated health management. However, more and more, these technical and programmatic issues are being addressed within the health management community.

4. USER OBJECTIVES AND METRICS

In order to present the ISHM user objectives and metrics we have chosen to broadly categorize types of users. There are many different ways to categorize these health management stakeholders. Our approach is shown in Figure 1.

Our three top-level stakeholder categories are Operations, Regulatory and Engineering. In this paper, we focus on looking at the user objectives derived from operations and how these impact the modeling efforts of the engineering activities.

The three distinct user groups consist of operations, regulatory and engineering. Within operations we have logistics, flight, maintenance, fleet management and training. Regulatory users are concerned mainly with establishing FAA amendments and new rules taking advantage of health management information. Within engineering we have sustaining, R&D and manufacturing. Although design engineers can be considered users of health management, due to space considerations we do not survey engineering design.

In the remainder of this article we have chosen to highlight each identified user objective only once even if it may be attributable to multiple users. For example, reducing labor is an objective that spans multiple users but the associated user metric is universal - hours of labor. Our categorization also has forced boundaries between user groups that may cause some of the objectives to be split. For example, one of the user objectives for logistics is to reduce the mean time to repair. We have chosen to put this under logistics rather than under maintenance as in (Hess et al., 2005).

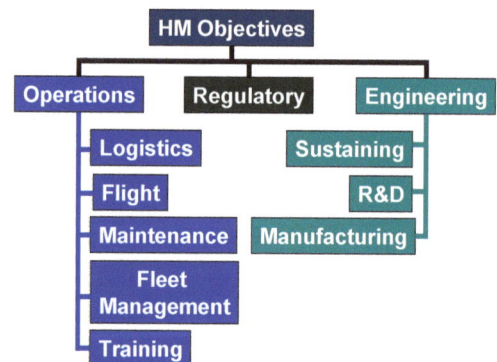

Figure 1. Categorization of groups driving health management objectives.

**Table 1. Logistics HM Goals and Metrics
(d: diagnostics, p: prognostics)**

Logistics Goals	User Metrics	Map
L.1 Reduce repair turn-around time	Mean time to repair (MTTR), time delays waiting for parts	d, p
L.2 Reduce ground support equipment and personnel	Equipment value, volume, weight and number of personnel	d
L.3 Increase availability/decrease unscheduled maintenance	Mean time in service	p
L.4 Reduce labor	Labor-hours	d, p
L.5 Reduce periodic inspections	Frequency of periodic inspections	p
L.6 Predict remaining useful life in components, maximize component life usage and tracking	Accuracy in prediction, minimize false alarms	p
L.7 CBM - Schedule regular maintenance only as necessary - Predict remaining useful life in expendables (e.g. oil)	Prediction accuracy	p
L.8 Ease of using entire information system	Measure of integration and information access: data access, security, search, increase IS availability, decrease costs…	d, p
L.9 Increase surge capacities	Surge capacity	d, p
L.10 Reduce costs of reconfigurations and turn-arounds	Total $ spent on reconfigurations	d, p
L.11 Maximize vendor lead time	Lead time	p
L.12 Inventory	Spare parts usage	d, p

4.1 Logistics

DEFINITION: Logistics is the science of planning and executing the acquisition, movement and maintenance of resources necessary to sustain aeronautical operations.

The bottom line for logistics is to make operations faster, cheaper (less stuff, less personnel) and more consistent and reliable (less uncertainty and more predictable). This top floor view of logistics can be translated into the user objectives and associated metrics listed in Table 1. All of the user objectives tables presented will have the rightmost column indicating whether the performance metric can be mapped into diagnostics (d), prognostics (p), both or neither.

In this table are objectives that would exist even without any health management solution such as reducing turn-around and repair times. Hopefully, these can be improved through the appropriate application of health management information. The issue of reducing ground support also exists whether or not we have a health management system. The central concept is that the diagnostic (fault type and location) information available will reduce the need for extensive ground test equipment and will reduce the time spent on facilitating repairs as well.

Reducing the frequency of periodic inspections by relying upon more extensive system monitoring is starting to become a reality in the Air Force [16]. The individual aircraft tracking program enables the development of an individualized aircraft specific maintenance schedule (including inspections) based on actual fatigue loads and/or crack lengths for each aircraft.

Without IVHM, consumables (such as oil) are replaced at a fixed schedule based upon expected usage. Condition based maintenance (CBM) [6, 7] has started using the operating regime to modify this replacement schedule and the inspection intervals. Heavy use will result in more frequent inspections and vice versa. Additionally, the actual condition of the consumable/expendable can be monitored either directly or indirectly based upon operating conditions. The rate of deterioration can be estimated and then the optimal replacement schedule predicted so that the operator can be notified in advance. This type of technology enables logistics to schedule service in advance at an optimal replacement schedule.

A final point on logistics is the user objective for ease of use of the entire information system (IS). This includes ensuring that the appropriate people/teams have access to the appropriate information at the right time with sufficient data integrity and security. Unfortunately, many times the information system is thought of after the fact as merely a way to archive records. This lack of integration has been identified as a large reason for failure (Hess and Fila, 2002). It should be noted that measuring "ease of use" for an entire IS is very difficult and subject to multiple, sometimes conflicting ideologies. Many measures associated with evaluating the usability of an enterprise system are subjective.

The Air Force has set the objective of modernizing the information systems that underlie its logistics with the goal to increase IS equipment availability by 20% and reduce annual operational expenses by 10% (Alford, 2007).

The objectives and metrics associated with an information system that spans all aspects of aviation operations are far beyond the scope of this article. However, we will highlight some of the key aspects with respect to health management and how a user might assess:

- asset tracking

- individual aircraft condition assessment

- demand management

- lifecycle product management

- integrated planning system

- purchasing supply chain management

- fleet decision management tools

The Joint Strike Fighter program is developing autonomic logistics information system tools to integrate management systems (e.g. fleet and maintenance) along with knowledge discovery tools and anomaly and failure resolution systems. Since the IS is responsible for enabling real-time information flow between maintenance, training, supply and mission planners as well as to provide data for performance analytics it can be considered the backbone of logistics (Byer et al., 2001).

In the past, such large scale integrated IS implementations have failed for a number of reasons such as poor understanding of the requirements, immature products, limited testing in actual environments and under appreciating and under valuing the effort required for data cleanup (Alford, 2007). Typically data useful for analytical modeling is contained in multiple heterogeneous systems. (Wilmering, 2003)

One difficulty of accurate maintenance data collection is more than just an information system issue - humans are the ones performing the maintenance actions and entering the maintenance data into the information system. In the past, the maintenance codes provided to maintenance technicians in both military and civilian sectors were rather coarse grained to enable easier entry during maintenance. This meant that during unscheduled maintenance debugging activities, there could be inaccuracies generated either from entering the closest (or most familiar) maintenance code or entering the wrong

premature diagnosis. For example, electrical wiring in the past was not considered as a separate system but rather was just the thing between reportable sub-systems. This meant that wiring problems were often under reported within the maintenance database. This has been addressed by adding additional maintenance codes and making the definitions more precise with the adverse consequences of requiring even more labor and costing more time for maintenance technicians.

Another aspect to ensuring the utility of the information system is through the use of common architectures, interoperability metrics, common standards and a clear path to implementation [29]. The U.S. Department of Defense (DoD) Architectural Framework (DoDAF) defines a standard way to organize an enterprise architecture into consistent views (DOD, 2007). Other approaches include ontological interchange standards KIF (KIF, 1998), product data oriented standards such as STEP (S. 1030-1, 1994), and even more specific diagnostic information models such as AI-ESTATE (Sheppard and Kaufmann, 1999).

In a similar vein, IEEE is also developing standards such as the Automatic Test Markup Language (S.1671.4, 2008) and the Software Interface for Maintenance Information Collection and Analysis (Sheppard and Wilmering, 2006) as a means to standardize the exchange of test information between automatic test equipment.

From a lessons learned perspective on the JSF program, a well integrated information system has been identified as the most important lesson learned (Hess and Fila, 2002). This lesson includes ensuring that ground systems are developed jointly with diagnostic systems and that on-board diagnostic algorithms are developed in a manner to ensure full system capability.

There is a great difference between supply chain management for a large operation consisting of a uniform fleet and managing a very small number of highly unique and complex vehicles (such as NASA's Shuttle Orbiter program). With a small number of vehicles requiring custom part specifications, the lead time to the vendors needs to be maximized, and having an inventory of such spare parts is advisable. In the case of large fleets where multiple sources are available for parts and supplies, a just in time inventory approach can help minimize waste and storage expenses. Turn-around time can be optimized through proper planning and use of analytical and prediction capabilities of fleets.

4.2 Flight

DEFINITION: The Flight category includes the pilots and flight crew as well as those responsible for Safety of Flight.

The bottom line for flight objectives for health management systems is to only provide information that increases certainty for future actions and commands and increases safety of flight.

Table 2 list objectives related to flight. A clear violation of the information certainty objective is false alarms - alerting the crew to a problem in a subsystem when the problem does not really exist.

A second objective, also related to reducing uncertainty in the cockpit, is the objective not to have conflicting alarms - also known as dissonance (Pritchett et al., 2002; Song and Kuchar, 2003). This objective can unfortunately be derived from a lessons learned from a tragic flight accident. In July 2002 a mid-air collision occurred between a Russian passenger jet and a DHL cargo jet over Germany which resulted in 71 deaths. Analysis of this accident revealed a dissonance problem between an on-board alerting system called the Traffic Alert and Collision Avoidance System and an air traffic controller whereby the Traffic Collision Avoidance System commanded the pilot to gain altitude to avoid a collision and the control tower commanded a decrease in altitude. The conflicting signals, even if the pilot can prioritize, cause time delays in executing the appropriate action.

Table 2. Flight User Goals and Metrics
(d: diagnostics, p: prognostics)

Flight Goals	User Metrics	Map
F.1 Minimize cockpit false alarm rate	Time between false alarms	d, p
F.2 Minimize cockpit information overload	# health management messages	d, p
F.3 Enable cockpit information filtering of critical alarms	Capability to filter - pilot satisfaction	d, p
F.4 Enable cockpit information filtering of non-critical alarms	Capability to filter - pilot satisfaction	d, p
F.5 Minimize alarm conflicts	# conflicting alarms	d, p
F.6 Minimize alarm dissonance	# alarms that have disparity between ATC and alarms	d, p
F.7 Maximize time from first alert to failure.	Time to failure or when safe landing becomes difficult.	d, p
F.8 Enhance Safety	# aborted flights	d, p
F.9 Enhance Safety	# smoke events	d, p
F.10 Enhance Safety	Passenger comfort complaint rate	d, p

The flight crew also needs to have as much advanced knowledge of an imminent failure as practical (Vincent and Pritchett, 2001). In particular, pilots need to be alerted early enough that the fault can be resolved and control regained (if lost) or if the handling qualities are too severely degraded, the health management system should be able to augment vehicle control stability in conjunction with a damage adaptive controller to enable a safe emergency landing.

The ability for crew to prioritize, although essential, is one that is easily overloaded when either too much information is presented or when the most critical information is either buried beneath layers of information or is not easily accessible (multiple sub-menus). This relates to both optimizing the number of health management messages sent to the crew as well as allowing for the crew to appropriately filter the less critical messages.

There is a lack of understanding in the community as to "how good is good enough" and "how good can we get" with respect to fault diagnosis. This is intimately connected with practical issues such as performance metrics and false alarm rates. In the past, on-board diagnostic systems have had a terrible record for costing more then was saved. For example, in Bain and Orwig (2000) it is documented that built-in-tests (BIT) caused wasted (CND) maintenance of the order of 85,639 maintenance man hours and 25,881 hours unnecessary aircraft downtime. This issue has plagued the F/A-18E/F and the V-22 Osprey (Westervelt, 2006).

Of course the top priority of the flight crew is safety. Typically safety can be measured in terms of the number of aborted flights, number of National Transportation Safety Board (NTSB) incident and accident reports, number of smoke events (when the smell or sight of smoke is present), and number of passenger comfort complaints (air quality, water quality, temperature…).

As the safety of air transportation continues to improve, the impact of health management systems on safety becomes increasingly difficult to measure. Nevertheless, the introduction of health management technology should always be required to improve safety. There is always risk from the introduction of technology that needs to be weighed and mitigated so that safety margins are always improving.

4.3 Maintenance

DEFINITION: Maintenance health management users are defined as the personnel in the depots and on the field responsible for repairing and servicing the aircraft.

**Table 3. Maintenance User Goals and Metrics
(d: diagnostics, p: prognostics)**

Maintenance Goals	User Metrics	Map
M.1 Decrease incidents of cannot duplicate (CND) logs and retests OK (RTOK)	# CNDs	d
M.2 Reduce failures	MTBF	p
M.3 Increase operation after non-critical faults	Time of operation after non-critical fault	p
M.4 Reduce damage incurred	# damage incidents logged as caused by maintenance	p
M.5 Reduce maintenance look-up time	Time to access maintenance manuals and records	d
M.6 Identify fault location	distance to fault in wiring, LRC identification	d
M.7 Reduce health management system maintenance	Hours spent on diagnosing and repairing the health management system	d, p
M.8. Maximize fault coverage	Percentage of detectable faults	d

**Table 4. Fleet Management Goals and Metrics
(d: diagnostics, p: prognostics)**

Fleet Management Goals	User Metrics	Map
FM.1 Life extension - in service beyond expected service life	Years past retirement	p
FM.2 Decrease unscheduled maintenance	Hours of unscheduled maintenance	p
FM.3 Easily reconfigurable	Time to respond to mission change	
FM.4 Efficiency	Fuel used vs. cargo/people transported	d, p
FM.5 Vehicle targeted CBM	(HUMS examples)	d, p
FM.6 Decrease ops costs (RMO)	operating expenses	p
FM.7 Increase availability	mean turn-around time	d, p
FM.8 Provide surge capacity	surge capacity	p
FM.9 Spare part usage analytics.	Percent accuracy on part usage predictions.	p
FM.10 Aid business and regulatory decisions		d, p
FM.11 Improve design and qualifications		d, p

The bottom line for maintenance is to as quickly and as inexpensively as possible return an aircraft to service while minimizing repeated repairs.

The most costly and most time consuming type of faults are intermittent faults seen during flight that cannot be duplicated (CND) in the maintenance depot. These faults may not be discovered by static depot tests. The dynamic environment of flight may cause faults which only manifest in flight. These types of faults result in subsystems (e.g. Line Replaceable Units - LRUs) being pulled for testing unnecessarily resulting in excessive inventory of parts that retest OK (ROK), excessive time spent on testing and trying to diagnose LRUs that actually are not faulty, and test flights trying to pin down the correct diagnosis. Health management systems hold the allure that a correct diagnosis (fault type and location) can be provided without intervention by the maintenance personnel. This would both reduce the incidents of CNDs and RTOKs as well as reduce the required labor. Table 3 contains a sampling of the maintenance objectives.

One of the greatest sources of faults for Electrical Wiring and Interconnect Systems comes from poor maintenance practices (Collins and Edwards, 2005). For example, if a new wire needs to be run, rather than unscrew the wire clamps and undo the wire ties along the harness it is quicker to just push the wire through the clamps and ties if it will fit. This can have the consequence of causing the wire clamps to be too tight resulting in pinching of all of the wires. Over time this can result in abrasion and breakage internal to a wire. Although this problem does not manifest right away, bad practices such as this can reduce the average fleet mean-time between failure (MTBF) values.

A prognostic capability within a health management system provides the capability to predict and trend degradation before eventual failure occurs. The ability for maintenance to reduce subsystem failures by repair and/or replacement prior to failure can be measured in terms of mean time between failures. In the case of the electrical wiring issue, a future electrical diagnostic system could sense the abnormal wear within the pinched wires. The prognostic system could then form an estimate as to when the stressed wires would need to be replaced to avoid interruption in service.

An ISHM system also holds the promise to reduce maintenance turn-round time by identifying the location of the fault. For discrete state based systems, the fault coverage can be extensive and enumerated. Fault coverage for analog parameters is much more difficult to ensure than with discrete domains. Typically due to the continuous nature of the range of parametric faults along with the inherent masking effect of process variations there tends to be a range of faults in which are not entirely detectable. This grows worse as variance increases. Two novel test metrics are introduced in Abderrahman et al. (2007) a guaranteed parameter fault coverage obtained by a deterministic method, which is the guaranteed lower bound of PFC, and a partial parameter fault coverage, which is the probabilistic component of PFC. The details of these metrics can be found in Abderrahman et al. (2007).

4.4 Fleet Management

DEFINITION: Fleet management health management users are defined as those involved with making fleet wide decisions affecting life extension, operational costs (RMO) and future planning.

The bottom line for fleet management is to maximize adaptability, availability and mission success while minimizing costs and resource usage.

Fleet managers interact with the health management system at a higher level of abstraction than do the other users. The accuracy of the analytics and system assessments is even more critical at this level due to the large consequence of a single misinformed decision. Since fleet management is at such a high level it encompasses the users that we have previously examined such as logistics, flight and maintenance. Table 4 summaries the objectives of fleet management.

Integral to fleet management is the use of decision support systems within an integrated information system. Decision support systems aid business and regulator decisions and improve design and qualifications by emphasizing specific query, reporting and analysis capabilities [44]. This is used both by a fleet owner and operator as well as by original equipment manufacturers (OEMs) (e.g. warranty calculations). This also impacts regulatory affairs by allowing fleet managers to have the information necessary to adhere to strict regulatory inspection intervals and minimize fleet wide disruptions.

5. DIAGNOSTIC&PROGNOSTIC SYSTEMS

The health management system user objectives and metrics will next be related to those metrics associated with development and operation of diagnostic and prognostic systems. Many of these user objectives will map into both diagnostic and prognostic metrics, others will not map into either. An extensive survey on

diagnostic metrics was conducted in Kurtoglu et al. (2008). The primary results from this survey are presented in Table 5. Surveys on prognostic metrics, including suggestions as to new metrics for prognostics are presented in (Saxena et al., 2008) and (Laeo et al., 2008). Readers wishing for more insight into performance measures for diagnostics and prognostics are directed to look at Kurtoglu et al., (2008), Szxena et al., (2008), Laeo et al., (2008) and the references contained therein.

5.1 Diagnostics

DEFINITION: Diagnosis is the detection and determination of the root cause of a symptom.

The bottom line for diagnostics is to detect and isolate faults in a timely and accurate manner with sufficient resolution so as to identify the specific faulty component.

The objectives and associated metrics for diagnostics taken from [1] are summarized in Table 5. The diagnostic objectives have been categorized into two categories: detect and isolate. Within each of these categories are objectives related to response time, accuracy, sensitivity/resolution and robustness. The previously presented user objectives for logistics, flight, maintenance, fleet management and training can be related to the diagnosis objectives and metrics in Table 5. A summary of this mapping of user goals to diagnostics is summarized in Table 6. The purpose of this mapping is to present the relationship between published user objectives and the performance measures used to drive diagnostic algorithm research and development.

In order to make this table presentable, we have selected the most important diagnostic measure(s)

**Table 5. Diagnostic Metrics
(Kurtoglu et al., 2008)**

Type	Diagnostic Objectives	Model Metrics
Detect	Time	Response time to detect
	Accuracy	Detection false positive rate
	Accuracy	Detection false negative rate
	Accuracy	Fault detection rate
	Accuracy	Fault detection accuracy
	Sensitivity	Detection sensitivity factor
	Stability	Detection stability factor
Isolate	Time	Time to isolate
	Time	Time to estimate
	Accuracy	Isolation classification rate
	Accuracy	Isolation misclassification rate
	Resolution	Size of isolation set
	Stability	Isolation stability factor

for each objective. This ignores many of the points that have been made within this article and should only be considered within that context. For example with MTTR, we have listed accuracy and specificity - this is not to say that timeliness is not important - timeliness is essential as has been pointed out within our discussion.

Note that there are several categories that have been listed as not defined. For example, the ease of using an IS is not clearly defined within the diagnostics development community. One of the user objectives: minimizing alarm dissonance - requires a more systems level approach than can be provided by listing a single diagnostic.

1) Diagnostics for logistics

All of the measures in Table 5 directly or indirectly impact some of the previously described user metrics. The logistics user goals and metrics (Table 1) relevant to diagnostics are related to the appropriate diagnostic metrics.

Reduce repair turn-around time (L.1) - The user goal of reducing repair turn-around time as measured by the mean time to repair can be facilitated via maximizing the accuracy of fault detection and isolation. An automated diagnostic system that can pin-point the fault type and faulty sub-system component will save technicians time in locating the root cause of the fault symptom. The reduction in repair turn-around time corresponds to the ratio between the time spent diagnosing with respect to the total time of diagnosing and repair. Conversely, a bad diagnosis system will mislead repair personnel and potentially adversely impact the repair turn-around time. Occasionally such misdirections will occur, it is important to evaluate the mean of the reduction in repair turn-around. If the deviation is too high about this mean the repair personnel may stop using the system out of frustration.

Reduce ground support equipment and personnel (L.2) - The goal of reducing ground support/footprint as measured by the number of ground support personnel and also by the amount of equipment required to diagnose a fault is related to all the entries of Table 5. If the diagnosis system is quick enough to transmit logistics requests prior to landing, and if the diagnosis is accurate and has high enough specificity (resolution), then right test equipment at the right time may be made available via on-board diagnostics telecasting the appropriate information to maintenance and logistics.

Reduce labor (L.4) - Reducing labor as measured in aggregate labor hours is enabled by ensuring accurate detection and isolation diagnosis as well as a timely solution. If the detection and isolation algorithms take longer to find a solution than the nominal labor required to discover root cause, then the system is a failure. Additionally, if the diagnosis or isolation is wrong too many times, technicians will spend more time to enact repairs and will eventually terminate usage of the diagnostic system.

An independent technical assessment of software for the F-22 determined that the acquisition activity failed to properly interpret and implement fault detection and fault isolation requirements (Marz, 2005). In particular, the following software capability gaps in the integrated diagnostics were highlighted:

1. Test coverage

2. Correlating faults to failures

   a. ability to isolate failures

   b. determining the consequence of a failure

3. Fraction of false alarms / false positives

4. Software health management - diagnostic environments that monitor software for faults are immature.

Test coverage refers to how many of the physical system failure modes are included within the scope of the diagnosis algorithms. The size of the isolation set (Table 5) refers to how many modes within the model scope are reported in a candidate set (size of the ambiguity group).

Ease of using entire information system (L.8) - The ease of use of the information systems associated with all aspects of the life-cycle is very difficult to measure and has many different meanings. For our purposes, we will relate this to the diagnostic objective of minimizing time to respond. It is very difficult and subjective to measure the performance of an information system from user perspectives. For example, the information needs, access rights and even processing operations vary greatly from logistics, to maintenance and fleet management. Fleet managers may need annualized aggregated statistics whereas maintenance personnel need access to an individual vehicle's repair history and to OEM part replacement procedures.

Minimize inventory (just in time) (L.12) - One of the ways to reduce the need for a large inventory of spare parts is to have a method by which repairs are initiated such that only those parts which need replacement are swapped out. Often times during a diagnostic procedure, a technician will need to swap out parts to try to localize the root cause of the fault. With a diagnostic system capable of accurate fault isolation this behavior of part swapping can be reduced thus impacting the inventory metric. Prognostics can have an even greater impact on minimizing required inventory by predicting wear trends.

Table 6. Diagnostic Mapping Summary

User Community Goals/Metrics	Diagnostic Saliency
Logistics Min. MTTR	Max. accuracy & specificity
Min. ground support	Max. specificity
Min. labor hours	Max. accuracy & specificity
Ease of use of IS	Not defined
Minimize inventory	Max. accuracy & isolation
Flight Min. false alarms	Max. accuracy
Min. info overload	Max. accuracy & specificity
Enable info filtering	Max. specificity
Min. alarm conflicts	Max. accuracy
Min. alarm dissonance	System level issue
Max. alert time from failure	Timeliness
Max. safety	All
Maintenance Min. CND & RTOK	Accuracy & isolation
Reduce look-up time	Not defined
Accurate fault location	Max. isolation and distance to fault
Min. HMS maintenance	Not defined
Max. fault coverage	Max. coverage
Fleet Max. efficiency	Accuracy & specificity
Vehicle CBM	Accuracy
Aid business decisions	Not defined
Improve design	Not defined

2) Diagnostics for flight

Automated diagnostics for the flight crew has more critical factors with respect to timeliness of reporting than logistics requires. The crew needs enough time to be able to either resolve the fault condition or to respond and plan for an emergency landing. Another metric within Table 5 that pertains to diagnostics for the flight crew is the measure of the number of false alarms.

Minimize cockpit false alarm rate (F.1) - The minimization of cockpit false alarms as measured by the time between false alarms is obviously mapped directly to the detection false positive rates. The metric of time between false alarms is not necessarily the optimal measure, not all false alarms will be treated equally by the crew. There is a measure of criticality that needs to be added to this metric.

Minimize cockpit information overload (F.2) - Information overload can cause crew to miss critical messages as well as to create patterns of behavior whereby ignoring messages is rewarded due to misinformation. In part this can be alleviated by improving the accuracy and specificity of the provided diagnostic information. Additionally, since different user preferences will prevail, there needs to be information filtering capabilities.

Enable cockpit information filtering of critical alarms as measured by pilot's satisfaction (F.3, F.4) - The capability to filter critical cockpit alarms can be measured by surveying pilot satisfaction. Whenever a metric involves measuring human satisfaction, the complexities can be enormous. The ability to filter messages can be considered independent from the diagnostic system as long as inaccuracies are mitigated. It is often the case that human factor issues are not adequately considered when diagnostics are developed at the sub-system level. These human centric issues become more apparent at a systems integration level.

Minimize alarm conflicts as measured by number of conflicting alarms and minimize alarm dissonance as measured by number of alarms that have disparity (F.5, F.6) - The number of conflicting alarms and the number of alarms that have disparity can be indicators of overall system integration. Many times diagnostics are developed independently for sub-systems by different vendors and then the central diagnostic system is responsible for reconciliation of all of those systems. The conflicts definitely arise from the error statistics of the individual sub-systems but there is a higher level of functionality that is not represented individually. The performance of minimizing conflicts can be measured by the accuracy and resolution of the integrated system. Alarm conflicts may involve dissonant information creating conflict between the control tower and the advisory cockpit warnings.

Even in the absence of control tower communications, cognitive dissonance resulting from alarms may cause a loss of situational awareness among the crew members and lead to incorrect actions being taken. This level of system integration is typically beyond scope of the majority of diagnostic and prognostic engineers.

Maximize time from first alert to failure as measured by time to failure or when landing becomes difficult (F.7) - Maximizing the in-flight timeliness of a diagnostic is critical to both giving the flight crew adequate time to plan and respond as well as giving the ground logistics time to implement a maintenance plan. Typically there is a trade-off between early detection and false alarms. It is frequently the case that early detection can only be made when more false alarms are allowed to be incurred. This trade-space needs to be weighed carefully with respect to the criticality of the failure and the amount of time really required to prepare for a safe landing.

Safety as measured by number of incidents (e.g. smoke events) or number of aborted flights (F.8-F.10). - Obviously underlying all improvements in all of the other categories is the need to always be maintaining or improving safety margins. All aspects of diagnostics relate to safety.

3) Diagnostics for maintenance

Decrease incidents of cannot duplicate (CND) logs and retests OK (RTOK) (M.1) - The maintenance health management users metric for the number of CND logs will be positively impacted by accurate fault detection and isolation.

Reduce maintenance look-up time (M.5) - Legacy systems can make even the simplest task take considerable time. For example, repairing a broken sensor wire requires that the maintenance personnel be able to lookup that particular sub-system in the OEM manuals to determine the wire type, correct size and routing. This information can be buried in obscure encodings and difficult to use manuals that are not readily accessible electronically in the maintenance bay. As diagnostic systems become more sophisticated, it is important that they make the necessary information immediately available to those personnel that will facilitate the repair.

Fault location (M.6) - Fault location is a bit trickier to map directly to Table 5, which lists fault isolation. Fault isolation in some sense implies more a discrete state-space approach. There are certainly subsystems such as electrical wiring, wherein both fault isolation and fault localization are different. For example, fault isolation determines which wire or wire bundle (or connector) is responsible for the given fault symptoms; whereas fault localization specifies the precise location (distance to fault) of the damage on the wire responsible for the fault. This will become an increasingly important distinction as arc fault circuit breakers come into operation. An arc fault circuit breaker is designed to interrupt the circuit once an arcing condition has been detected. Unfortunately, by the time arcing has been detected, there will be damage present on at least one wire. This damage will typically be just a small spot (a consequence of an effective breaker) and may be very difficult to find via visual inspection without location information.

Health management system maintenance (M.7) - Another aspect that is unique to diagnosis is the maintenance required to maintain the health of the diagnostic health management system. Although this does not appear in Table 5, the maintenance objectives for the diagnostic health management system need to be one of the factors within the model metrics. It is important that such issues as sensor fatigue/failure be diagnosed appropriately rather than misclassified as a fault with the system that the sensor(s) is measuring. Although time and money savings will be incurred through a healthy health management system, if the maintenance of the HMS consumes all of these savings then a net result has been to increase risk to safe operation of the vehicle. Another application where fault localization is of great importance is structural health management.

Fault coverage (M.8) - Fault coverage for discrete fault states indicates the percentage of faults that the diagnosis system is able to detect and diagnose. It is important that the fault coverage includes the health management system itself so that technicians are better able to direct their attention to the appropriate sub-system. For continuous fault states, the coverage indicates the ability to detect faults within acceptable limits. Fault coverage is impacted by the resolution of the diagnostic system. A system that has broad coverage but is not able to localize will not have much of an impact on turn-around time. This is also related to the isolation set which determines the resolution of the diagnoses.

4) Diagnostics for fleet management

Diagnostics for fleet management has the potential to reduce the number of maintenance hours and thereby positively impact the user metrics of mean turn-around time and hours of unscheduled maintenance, although the number of maintenance activities will not likely decrease. Ultimately, the other fleet user objectives and additionally the unscheduled maintenance metric will be impacted by an effective prognostic system.

Efficiency (FM.4) - All systems on a vehicle may be running within nominal operating ranges but peak efficiency may not be achieved when some systems are near the edge of nominal behavior. The ability to trend these in prognostics will have the greatest impact on improving and maintaining optimal performance efficiencies.

Vehicle targeted CBM (FM.5) - Condition based maintenance, with sufficient information technology infrastructure, can be targeted to individual vehicles making it possible to optimally maintain a vehicle based upon its history as well as operating context. An accurate and specific diagnosis system integrated within a larger information system can enable vehicle targeted CBM.

Increase availability (FM.7) - Diagnostics can aid in increasing average fleet availability by minimizing the mean time to repair (by providing accurate diagnoses). Prognostics will have an even greater impact by minimizing the down time attributable to unscheduled maintenance fleetwide.

Aid business and regulatory decisions (FM.10) - Well thought out system integration is essential for diagnostics to be able to impact business decisions. For example as a fleet ages, vehicles start to exceed the original expected life, failures may start to be diagnosed in a few vehicles that are both the source of unscheduled maintenance as well as indicative of a bad trend. These fleet wide diagnosis trends can be analyzed to determine when is the optimal time to schedule replacement of parts in the non-failed part of the fleet prior to failure but without prognostics or trending degradation.

Improve design and qualifications (FM.11) - As parts are diagnosed as failing, there may be fleet wide occurrences of component failures that were not expected by the engineers. A diagnostic system that is well integrated into a fleet wide information system can alert engineering that an analysis needs to be performed to determine if these components will continue to fail at an unexpected rate thus warranting a design improvement.

5.2 Prognostics

DEFINITION: Prognostics is defined as the ability to detect, isolate and diagnose mechanical and electrical faults in components as well as predict and trend the accurate remaining useful life (RUL) of those components (Banks and Merenich, 2007).

The bottom line for prognostics is to as accurately and as far in advance as possible predict the remaining useful life of components and consumables to aid in logistics management, maintenance planning, crew alerting (impending failure) and fleet-wide planning. From a maintenance perspective:

"The goal of the prognostics portion of PHM is to detect the early onset of failure conditions, monitor them until just prior to failure, and inform maintenance of impending failures with enough time to plan for them. This will, in effect, eliminate the need for many of the inspections, as well as provide enough of a lead time to schedule the maintenance at a convenient time and to order spare parts in advance." (Hess and Fila, 2007)

The prognostics discussion will for the most part not overlap with the previous discussion of diagnostics even though many of the points are strongly inter-related. Specifically, if a system cannot reliably detect a fault useful for diagnosis it will prove very difficult to accurately assess the remaining useful life of such a component. For diagnosis (Table 5) we broke the field into two types: detect and isolate. For each of these types we had measures for time, accuracy, sensitivity/resolution and stability. The model metrics for prognostics are taken from two overviews of performance metrics: (Saxena et al., 2008) and (Leao et al., 2008) as summarized in Table 7.

The detect category for prognosis has a different meaning from detect in diagnosis. As an example, consider the meaning of false positive for each. A false positive in diagnosis detection means that the diagnosis system detected and indicated a fault where none existed. However, a false positive in prognosis means that a prediction of failure is unacceptably early resulting in loss of usable service life. Thus, prognosis detection is with respect to a time horizon which depends on user requirements. Typically the notion of detection in diagnosis is not relative to a time horizon. For prognosis

**Table 7. Prognostic Model Metrics
(Saxena et al., 2008 and Leao et al., 2008)**

Type	Prognostic Objectives	Model Metrics
Detect	Accuracy of characterization	Early prediction, late prediction (with respect to time window)
	Missed estimation rate	# missed detections/total # prognoses
Predict	Accuracy of predict remaining useful life	Accuracy at specific times (error, average error)
	Minimize sensitivity	Sampling rate robustness
	Precision	Ratio of precision to horizon length, standard deviation
	Hit rate	# correct prognoses/total # of prognoses
	Timeliness	Prognostic horizon, accuracy at specific times, convergence rate
Isolate	Phase difference between samples and prediction.	Anomaly correlation coefficient
	Precise correct estimation rate	# correct prognoses without adequate resolution
Effectivity	Minimize number of required sensors	Reduced feature set robustness
	Minimize amount of data needed	Data frame size
	Prognosis effectivity	# avoided unsched. maint. events/total # of possible unsched. events for component
	Average bias	average wasted life of component

we have added two more types to detect and isolate: predict and effectivity. Similar to the diagnosis table, within these categories are objectives related to accuracy, time, sensitivity and effectiveness.

The last type: effectivity relates very much to engineering design trade-space. Designing a system to make diagnosis and prognosis easier is an extensive subject that is beyond the scope of this paper. However, the effectivity section is very much related to the cost benefits analysis discussion.

The metrics employed in prognostic algorithm research and development are shown in Table 7 as taken from (Saxena et al., 2008) and (Leao et al., 2008) The mapping between the user goals and the prognostic metrics are listed in Table 8. As was true with the mapping from user objectives to diagnosis, the purpose of this mapping is to present the relationship between published user objectives and the performance measures used to drive prognostic algorithm research and development. We will not repeat elements covered in the diagnosis discussion, we chose to highlight items specific to prognosis. The considerations and caveats implored for Table 6 apply to Table 8 as well. Higher level functions such as business analytics and decision support systems which are functions of prognostics have not been defined within the prognostics measures.

1) Prognostics for logistics

The logistics goals unique to prognosis from diagnosis are discussed hence.

Increase availability/decrease unscheduled maintenance (L.3) - Decreasing unscheduled maintenance (and therefore increasing availability) is directly enabled through accurate degradation trend prediction combined with adequate time horizons. The time horizon needs to be long enough to allow for proper scheduling of maintenance and logistics as well as to plan for the usage of replacement aircraft. Obviously, accuracy in prediction of remaining useful life is critical to not waste part life through premature replacement. Incorrect estimates are even worse in that they will result in replacement of good parts and unnecessary downtime or even worse - failed parts that would have otherwise been replaced before failure. The positive impact on availability of prognostics is great but the risk posed by inaccurate prognostics is equally great.

Reduce periodic inspections (L.5) - The U.S. Air Force is employing condition based maintenance techniques combined with usage predictions to change the frequency of inspections and replacements based upon usage. The impact on the commercial sector with such technology depends largely on regulatory affairs (FAA). The current regulations need to take into account the ability to monitor and predict degradation trends as a means to reduce periodic inspections. Accuracy of the predictions of remaining useful is essential to avoid replacing parts that still have good life left and to avoid unscheduled maintenance due to unpredicted failures.

Predict remaining useful life in components, maximize component life usage and tracking (L.6) - Maximizing component life usage means having the ability to accurately know when an isolated component will fail with enough lead time so as to be able to schedule replacement.

Obviously this relies upon accuracy of predictions as well as having an adequate time horizon and being able to isolate trends to specific components.

CBM - Schedule regular maintenance only as necessary - Predict remaining useful life in expendables (e.g., oil) (L.7) - One of the first applications of prognostics has been in assessing the state of consumables such as oil. Oil can be monitored for its quality, for contaminants, and for quantity. The trend of the degradation of the oil can then be predicted and used to optimally schedule maintenance for replacement/renewal.

Provide surge capacity (L.9) - The ability to delay or adjust maintenance windows provides the capability of supporting surges in operations. Accurate health predictions aid in understanding the limits to possible delays and adjustments.

Reduce costs of reconfigurations and turn-arounds (L.10) - Unusual or unanticipated maintenance problems can result in costly reconfigurations of the supply chain or interruption of typical logistics processes. The ability to quickly and accurately identify the causes of faults and predict failures results in less disruption to establish procedures and protocols, thereby saving time and money. Additionally, planned reconfigurations can be scheduled to incorporate preventive maintenance that might not otherwise have occurred if not for accurate predictions of remaining useful life of components.

Maximize vendor lead time (L.11) - For parts that infrequently need to be replaced, but which require significant lead time for production and/or are expensive to maintain in inventory the ability to predict far in advance the trend in degradation is important. This relies upon have accuracy in prediction and isolation with enough time horizon to facilitate logistics part ordering.

Minimize inventory (L.12) - Minimizing the required inventory by transitioning to a just-in-time inventory system requires both an adequate time horizon in the remaining useful life estimate as well as specificity so that there is enough time to order the correct parts.

2) Prognostics for flight

Although all of the elements of Table 2 have been covered in the discussion of diagnostics, we would like to discuss again one of the elements that can be positively impacted by prognostics.

Maximize time from first alert to failure (F.7) - Whereas diagnosis is responsible for detecting fault conditions - hopefully prior to full failure - prognosis is responsible for predicting the trend in degradation resulting in an estimate of the remaining useful life along with appropriate estimates of

Table 8. Prognostic Mapping Summary

User Community Goals/Metrics	Prognostic Saliency
Logistics Max. mean time in service	Max pred. accuracy & time
Max. surge capacity	Accuracy of predictions
Min. freq. of inspections	Accuracy of predictions
Predict life remaining	Accuracy, time, isolation
CBM	Accuracy
Max. vendor lead time	Accuracy and isolation
Minimize inventory	Accuracy and timeliness
Flight Max. alert time from failure	Time horizon, isolation
Maintenance Reduce failures - MTBF	Accurate prediction
Increase op after fault	Accurate trends and isolation
Reduce damage incurred	Accurate trending
Min. HMS maintenance	Effectivity
Fleet Max. life extension	Accuracy of predictions
Min. unscheduled maint	Accuracy & isolation
Min. RMO costs	Accuracy of predictions
Spares analytics	Not defined
Aid business decisions	Not defined

uncertainty (confidence bounds). This distinction for in-flight is critical marking the difference between, for example, stating that hydraulic pump is faulty versus warning that the performance of the hydraulic pump is trending downwards but will be operational for several more hours. Thus an accurate estimate of remaining useful life provides the flight crew with more options as well as providing logistics and maintenance more options for scheduling repairs.

3) Prognostics for maintenance

Reduce failures (M.2) - Currently when a component or sub-system fails that is not expected to fail the result is unscheduled maintenance and downtime for the vehicle. The hope of prognostics is that some of these unscheduled maintenance activities may be mitigated by trend prediction of degradation. With an accurate modeling of the trending, estimates of remaining useful life can be used to facilitate repair/replacement of components prior to failure thus resulting in reduced unscheduled maintenance. The trade-off is that if the prognostic system is overly conservative wasted component life will occur.

Increase operation after non-critical faults (M.3) - Some faults that have either been detected or are trending can be safely deferred for maintenance to avoid operation

interruption. It is critical that a detected fault is highly accurate and isolated and that the prediction of the trend also be highly accurate. The liability for the operator ignoring a fault due to misinformation from the prognostic system is high and every validation and redundant verification needs to be enacted.

Reduce damage incurred (M.4) - Electrical arc fault interruption circuit breakers are designed to augment traditional thermal based circuit breakers by monitoring for the electrical signature associated with arcing events and then cutting off the current flow to the arcing wire. The extension to this is to incorporate a chafing detection system to these breakers to be able to assess the state of wire insulation degradation with the aim of providing both a remaining useful life estimate as well as a distance to fault assessment. The ability to detect chafes prior to an arcing event allow for inspections and maintenance to be scheduled prior to damage occurring from arcing.

Of course underlying all of the remaining useful life (RUL) estimates produced by a prognostic system are stochastic processes. This revelation requires that a probabilistic sensitivity analysis be included as part of the validation and verification process of new prognostic systems (Kacprzynski et al., 2004). This also means that "saying a widget will fail in 100 hours is not sufficient. Saying that a widget will fail in 95 to 105 hours with 94 percent confidence is much more useful." (Line and Clements, 2006).

4) Prognostics for fleet management

Life extension - in service beyond expected service life (FM.1) - One of the consequences of operating a fleet beyond expected service life is an increasing in maintenance, both scheduled (more frequent) and unscheduled. The promise of prognostics is that the trend analyses can help mitigate unscheduled maintenance. More frequent maintenance may also be mitigated through careful monitoring if the appropriate regulatory authorities concur.

Decrease unscheduled maintenance (FM.2) - This is one of the greatest promises of prognostics, the ability to trend and predict the remaining useful life of a component prior to failure. The accuracy and specificity of such a prediction can enable maintenance to be performed as convenient but prior to failure. This should reduce the number of unscheduled maintenance occurrences.

Decrease ops costs (RMO) (FM.6) - A large factor in aging RMO costs stems from unscheduled maintenance. If this unscheduled maintenance can be mitigated with prognostics, then the RMO costs can be maintained at a more uniform level as the fleet age increases.

Provide surge capacity (FM.8) - The capability of being able to predict when a component will fail translates to the ability to schedule maintenance at a greater convenience. Greater flexibility in scheduling maintenance enables being able to provide planned surges in capacity.

Spare part usage analytics and business decisions (FM.9) - Usage and repair analytics are a higher level analysis function that require a highly integrated information system combined with the diagnostic and prognostic systems.

6. DISCUSSION

As a result of this survey we have found that although the metrics associated with diagnostic and prognostic algorithm and system performance will positively impact the user community, that there are gaps within the diagnostic and prognostic metrics. These gaps tend to fall within one of the following categories:

- Process

- System analysis

- Data management

- Verification and validation

- Human factors

Process - Large scale adoption of a fully integrated health management system requires buy-in from many different types of users as well as proper systems analysis methods to make the return on investment business case. Objectives and requirements should be generated from inclusion and ownership of a broad spectrum of users. Cost benefit analyses and education of users and management about benefits of health management help with the adoption.

It is possible for the best ideas in health management system development and operation to be foiled by archaic business policies. Cost savings ideas such as "Replace only on failure" as pointed out in ()Yukish et al., 2001) will result in a health management system showing no benefits. In other words, one of the biggest obstacles to health management system adoption is the undocumented human element. Buy-in must be obtained at all user levels for the successful adoption.

System analysis - Another large obstacle is the development of sophisticated technologies without a view to the greater system. This problem is often encountered with engineering development efforts devoted to sub-system diagnostics and prognostics. This system level perspective is often not considered by researchers in diagnostics and prognostics.

Each subsystem within diagnostics or prognostics can be engineered to successfully meet appropriate metrics but fail when verification and validation of the broader system

are considered. This means that a broader perspective on verification and validation of total system health management is needed where the whole system requirement is greater than the sum of the user requirements.

Data management - In addition to the system level perspective, there are integration issues, especially within the context of a broader information system, that will not be addressed directly by sub-system requirements or user requirements and yet will vastly impact perceptions of utility. Issues such as business analytics and decision systems are typically not directly considered with diagnostics and prognostic and yet are direct consumers of the information that is sourced from such systems.

Verification and validation - As OEMs start to outsource more subsystem developments, the total system validation and verification becomes a greater challenge. In particular, the V & V of complex interacting software systems would benefit from a model based verification approach as adopted by hardware developers. Ofstun (2002) also discusses that proper verification and validation of IVHM functionality cannot simply be verified in a laboratory, that incremental demonstrations need to be conducted and that after delivery anomalies will occur and the IVHM system needs to be easy to update.

Human factors - The human element is hard to quantify and easier to ignore than other performance metrics. In particular, issues such as alarm dissonance and conflicts derive from system wide activities not directly measured by any particular subsystem metric are hard to manage and mitigate.

7. SUMMARY

We have briefly surveyed the recent literature pertaining to user goals for aeronautic health management systems. We have compared these goals with the results from surveys of the objectives and metrics of diagnostic and prognostic method developments. Although many of the mappings have been concerned with diagnosis accuracy and isolation as well as the horizon and accuracy of prognosis prediction, some of the most interesting information is in the gap between user goals and the success metrics associated with diagnostics and prognostics.

NASA's Aviation Safety Program is investing in IVHM. NASA's IVHM project seeks to develop (http://www.aeronautics.nasa.gov/programs_avsafe. htm) validated tools, technologies, and techniques for automated detection, diagnosis and prognosis that enable mitigation of adverse events during flight. The project includes a systems analysis aspect that assesses i) future directions and technology trends in research related to detection, diagnosis, prognosis, and mitigation as they pertain to the stated goals of the IVHM project, and ii) requirements for future aircraft and the issues arising from current and near-term aviation technology. We note that while the primary focus of the IVHM project is on-board, the health management objectives discussed in this paper impact the entire aircraft life-cycle.

Other studies have developed lists of lessons learned with respect to aeronautic health management systems (Ofstun, 2002). In particular, (Novis and Powrie, 2006) points out that holistic approach which views the system as a whole rather than as a collection of parts is essential. This is also true regarding generating user requirements and garnering broad organizational support.

It is our hope that this survey of user objectives as well as the mapping from user objectives to diagnostic and prognostic metrics can further the widespread adoption of health management technologies within aeronautics.

ACKNOWLEDGMENT

We extend our gratitude to Joseph Totah (NASA), Richard Ross (NASA), and many others for valuable discussions in establishing the IVHM user requirements.

This research was supported in part by the National Aeronautics and Space Administration (NASA) Aeronautics Research Mission Directorate (ARMD) Aviation Safety Program (AvSP) Integrated Vehicle Management (IVHM) Project. Additionally, this material is based upon the work supported by NASA under award NNA08CG83C.

REFERENCES

10303-1, "Industrial automation systems and integration - Product data representation and exchange Part 1: Overview and fundamental principles." 1994.

I. 1636.1, "IEEE Trial-Use Standard for Software Interface for Maintenance Information Collection and Analysis (SIMICA): Exchanging Test Results and Session Information via the eXtensible Markup Language (XML)," February, 2008.

I. S. 1671.4, "IEEE Trial-Use Standard for Automatic Test Markup Language (ATML) for Exchanging Automatic Test Information via XML: Exchanging Test Configuration Information," April, 2008.

"727 to 787: Evolution of Aircraft Maintenance Systems," *Avionics Magazine Special Report:http://www.aviationtoday.com/Assets/Honeyw ellsmall.pdf,* 2007.

"DoD Architectural Framework: Vol. 1: Definitions and Guidelines, Vol. 2:Product Descriptions, Vol.3: Architecture Data Description," April, 2007.

Abderrahman A., M. Sawan, Y. Savaria, and A. Khouas, "New Analog Test Metrics Based on Probabilistic and Deterministic Combination Approaches," in *14th IEEE International Conference on Electronics, Circuits and Systems*, 2007, pp. 82-85.

Albert A.P., E. Antoniou, S. D. Leggiero, K. A. Tooman, and R. L. Veglio, "A Systems Engineering Approach to Integrated Structural Health Monitoring for Aging Aircraft," Wright-Patterson Air Force Base, Ohio, 2006.

Alford R., "Data Management as the Key to Prognostic Capability," in *SAE doD Maintenance Symposium \& Exhibition*, 2007.

Ashby M.J. and R. J. Byer, "An Approach for Conducting a Cost Benefit Analysis of Aircraft Engine Prognostics and Health management Functions," in *IEEE Aerospace Conference Proceedings.* vol. 6, 2002, pp. 2847-2856.

Bain K. and D. G. Orwig, "F/A-18E/F Built-in-test (BIT) Maturation Process," *National Defense Industrial Associated 3rd Annual systems Engineering & Supportability Conference,* October, 2000.

Banks J. and J. Merenich, "Cost Benefit Analysis for Asset Health Management Technology," in *IEEE Annual Reliability and Maintainability Symposium*, 2007, pp. 95-100.

Banks J., K. Reichard, E. Crow, and E. Nickell, "How Engineers Can Conduct Cost-Benefit Analysis for PHM Systems," in *IEEE Aerospace Conference*, 2005, pp. 3958-3967.

Boller C., "Ways and Options for Aircraft Structural Health Management," in *Smart Materials and Structures.* vol. 10, 2001, pp. 432-440.

Byer B., A. Hess, and L. Fila, "Writing a Convincing Cost Benefit Analysis to Substantiate Autonomic Logistics," in *IEEE Aerospace Conference.* vol. 6, 2001, pp. 3095-3103.

Collins J.H. and G. Edwards, "The Naval Air Systems Command's Initiatives for Aircraft Wiring Diagnostic Support Equipment for Multiple Maintenance Levels," *Naval Air Systems Command,* 2005.

Crow E., "Condition Based Maintenance, the Maintenance Execution Process and the Open Systems Standards, An Overview of Activities at Penn State," in *Defense Maintenance Conference* Reno, Nevada, 2006.

Hecht H., "Prognostics for Electronic Equipment: An Economic Perspective," in *IEEE Reliability and Maintainability Symposium*, 2006, pp. 165-168.

Hess A. and L. Fila, "Prognostics, from the Need to Reality - from the Fleet Users and PHM System Designer/Developers Perspectives," in *IEEE Aerospace Conference.* vol. 6, 2002, pp. 2791-2797.

Hess A., G. Calvello, and P. Frith, "Challenges, Issues, and Lessons Learned Chasing the Big P:

Real Predictive Prognostics Part 1," in *IEEE Aerospace Conference*, 2005, pp. 3610-3619.

Hess A., G. Calvello, and T. Dabney, "PHM a Key Enabler for the JSF Autonomic Logistics Support Concept," in *IEEE Aerospace Conference*, 2004, pp. 3543-3550.

Hoyle C., A. F. Mehr, I. Y. Tumer, and W. Chen, "Cost-Benefit Quantification of ISHM in Aerospace Systems," *ASME 2007 International Design Engineering Technical Conferences & Computers and Information in Engineering Conference,* September 4-7, 2007 2007.

Kacprzynski G.J., A. Liberson, A. Palladino, M. J. Roemer, A. J. Hess, and M. Begin, "Metrics and Development Tools for Prognostic Algorithms," in *IEEE Aerospace Conference*, 2004, pp. 3809-3815.

Kurtoglu T., O. J. Mengshoel, and S. Poll, "A Framework for Systematic Benchmarking of Monitoring and Diagnostic Systems," *IEEE International Conference on Prognostics and Health Management,* October 6-9, 2008.

Leao B.P., T. Yoneyama, G. C. Rocha, and K. T. Fitzgibbon, "Prognostics Performance Metrics and their Relation to Requirements, Design, Verification and Cost-Benefit," *IEEE International Conference on Prognostics and Health Management,* 2008.

Line J.K. and N. S. Clements, "Prognostics Usefulness Criteria," in *IEEE Aerospace Conference*, 2006.

MacConnell J.H., "ISHM & Design: A review of the benefits of the ideal ISHM system," in *IEEE Aerospace Conference*, 2007.

Marz T.F., "Integrated Diagnostics: Operational Missions, Diagnostic Types, Characteristics, and Capability Gaps," 2005.

Millar R.C., "A Systems Engineering Approach to PHM for Military Aircraft Propulsion Systems," in *IEEE Aerospace Conference*, 2007.

KIF, "Knowledge Interchange Format, Working draft of proposed American national standard," vol. Document No. X3T2/98-004, 1998.

Nenninger G., "Aviation Condition Based Maintenance (CBM)," in *DoD Maintenance Symposium \& Exhibition*, 2007.

Novis A. and H. Powrie, "PHM Sensor Implementation in the Real World - a Status Report," in *IEEE Aerospace Conference*, 2006.

Ofsthun S., "Integrated Vehicle Health Management for Aerospace Platforms," *IEEE Instrumentation & Measurement Magazine,* pp. 21-24, September, 2002 2002.

Pritchett A.R., B. Vandor, and K. Edwards, "Testing and Implementing cockpit Alerting systems," in *Reliability Engineering and System Safety.* vol. 75, 2002, pp. 193-206.

Saxena A., J. Celaya, E. Balaban, K. Goebel, B. Saha, S. Saha, and M. Schwabacher, "Metrics for Evaluating Performance of Prognostic Techniques," *IEEE International Conference on Prognostics and Health Management,* October 6-9, 2008.

Scandura P.A., "Integrated Vehicle health Management as a System Engineering Discipline," in *The 24th Digital Avionics Systems Conference.* vol. 2, 2005.

Sheppard J..W. and T. J. Wilmering, "Recent Advances in IEEE Standards for Diagnosis and Diagnostic Maturation," in *IEEE Aerospace Conference*, 2006.

Sheppard J.W. and M. A. Kaufman, "IEEE Information Modeling Standards for Test and Diagnosis," in *5th Annual Joint Aerospace Weapon System Support, Sensors and Simulation Symposium* San Diego, CA, 1999.

Song L. and J. K. Kuchar, "Dissonance Between Multiple Alerting Systems Part I: Modeling and Analysis," in *IEEE Transactions on Systems, man, and Cybernetics-Part A: Systems and Humans.* vol. 33, 2003.

Vincent B. and A. R. Pritchett, "Requirements Specification for Health Monitoring Systems Capable of Resolving flight Control System Faults," in *The 20th Conference Digital Avionics Systems.* vol. 1, 2001, pp. 3D4/1 - 3D4/8.

Westervelt K., "Fixing BIT on the V-22 Osprey," *IEEE Aerospace Conference,* March, 2006.

Wilmering T.J., "When Good Diagnostics Go Bad - Why Maturation is Still Hard," in *IEEE Aerospace Conference.* vol. 7, 2003, pp. 3137-3147.

Yukish M., C. Byington, and R. Campbell, "Issues in the Design and Optimization of Health Management Systems," in *New Frontiers in Integrated Diagnostics and Prognostics, Proceedings of the 55th Meeting of the Society for Machinery Failure Prevention Technology* Virginia Beach, Virginia, 2001.

Zhang S., R. Kang, X. He, and M. G. Pecht, "China's Efforts in Prognostics and Health Management," in *IEEE Transactions on Components and Packaging Technologies.* vol. 31, 2008.

Kevin R. Wheeler received the B.S. and M.S. degrees from the University of New Hampshire, Durham, in 1988 and 1991, respectively, and the Ph.D. degree from the University of Cincinnati, Cincinnati, OH, in 1996, all in electrical engineering. After graduation, he joined the IBM Almaden Research Center to develop web-mining algorithms. In 1997, he joined the Computational Sciences Division, NASA Ames Research Center, Moffett Field, CA. His research interests include applying probability theory to automating problems in integrated vehicle health management, the Earth sciences, man–machine interfaces, and robotics.

Tolga Kurtoglu is a Research Scientist with Palo Alto Research Center in the Embedded Reasoning Area. His research focuses on the development of prognostic and health management systems, model-based diagnosis, design automation and optimization, and risk and reliability based design. He received his Ph.D. in Mechanical Engineering from the University of Texas at Austin in 2007 and has an M.S. degree in the same field from Carnegie Mellon University. Dr. Kurtoglu has published over 40 articles and papers in various journals and conferences and is an active member of ASME, ASEE, AIAA, and AAAI. Prior to his work with PARC, he worked at MCT/NASA Ames Research Center as a research scientist and at Dell Corporation as a design engineer.

Scott Poll is a Research Engineer with the National Aeronautics and Space Administration (NASA) Ames Research Center, Moffett Field, CA, where he is the deputy lead for the Diagnostics and Prognostics Group in the Intelligent Systems Division. He is co-leading the evolution of a laboratory designed to enable the development, maturation, and benchmarking of diagnostic, prognostic, and decision technologies for system health management applications. He was previously the Associate Principal Investigator for Prognostics in the Integrated Vehicle Health Management Project in NASA's Aviation Safety Program. He received the BSE degree in Aerospace Engineering from the University of Michigan in 1994, and the MS degree in Aeronautical Engineering from the California Institute of Technology in 1995.

Author Index

Author Guidelines

The International Journal of Prognostics and Health Management (IJPHM) publishes scientific papers dealing with all aspects of prognostics, diagnostics, and system health management of complex engineered systems. High quality articles focused on assessing the current status and predicting the future condition of an engineered component and/or system of components. Such articles may come from a variety of disciplines, including electrical, electronics, mechanical, civil, and chemical engineering, computer and materials science, reliability, test and measurement, artificial intelligence, physics, and economics.

Copyright

The Prognostic and Health Management Society advocates open-access to scientific data and uses a Creative Commons license for publishing and distributing any papers. A Creative Commons license does not relinquish the author's copyright; rather it allows them to share some of their rights with any member of the public under certain conditions whilst enjoying full legal protection. By submitting an article to the International Conference of the Prognostics and Health Management Society, the authors agree to be bound by the associated terms and conditions including the following: As the author, you retain the copyright to your Work. By submitting your Work, you are granting anybody the right to copy, distribute and transmit your Work and to adapt your Work with proper attribution under the terms of the Creative Commons Attribution 3.0 United States license. You assign rights to the Prognostics and Health Management Society to publish and disseminate your Work through electronic and print media if it is accepted for publication. A license note citing the Creative Commons Attribution 3.0 United States License, as shown below, needs to be placed in the footnote on the first page of the article.

First Author et al. This is an open-access article distributed under the terms of the Creative Commons Attribution 3.0 United States License, which permits unrestricted use, distribution, and reproduction in any medium, provided the original author and source are credited.

Ethics

Contributions to IJPHM must report original research and will be subjected to review by referees at the discretion of the Editor. IJPHM considers only manuscripts that have not been published elsewhere (including at conferences), and that are not under consideration for publication or in press elsewhere. Moreover, it is the responsibility of the author to ensure that any data or information submitted complies with the export-control regulations of the author's home country (e.g., International Traffic in Arms Regulations (ITAR) in the United States). IJPHM honors code of conduct provided by the Committee of Publication Ethics (COPE). More details on IJPHM policies and publication ethics can be found online.

Submission Types

IJPHM publishes full-length regular papers, technical briefs, communications, and survey papers.

Full-Length Regular Papers should describe new and carefully confirmed findings, and experimental procedures and results should be given in detail sufficient for others to replicate the work. A full paper should be long enough to describe and interpret the work clearly, placing it in the context of other research.

Technical Briefs usually describe a single result, experiment, or technique of general interest for which a short treatment is appropriate. A short paper should be long enough to describe experimental procedures and clearly, and interpret the results in the context of other research.

Communications are a separate class of short manuscripts that are subject to an expedited review process. Appropriate items include (but are not limited to) rebuttals and/or counterexamples of previously published papers. A short communication is suitable for recording the results of complete small investigations or giving details of new models or hypotheses, innovative methods, techniques or apparatus. The style of main sections need not conform to that of full-length papers. Short communications are 2 to 4 printed pages in length. The Editors will review these submissions internally, and request outside review when appropriate.

Survey Papers covering emerging research topics in PHM are also published, and unsolicited manuscripts of a tutorial or review nature are welcome. However, prospective authors of survey papers should contact in advance the Editor-in-Chief in order to assess the possible interest of the topic to IJPHM. Papers describing specific current applications are encouraged, provided that the designs represent the best current practice, detailed characteristics and performance are included, and they are of general interest.

Prospective authors should note that for any type of IJPHM content, poorly documented papers using "proprietary" techniques will be rejected. Moreover, excessive "branding" within a paper also cause for rejection; e.g., "The team used the magical CompanyBrand™ preprocessing to prepare the data to extract the amazing CompanyBrand™-proprietary features (which we can't tell you about)." Papers should present techniques and results clearly and objectively.

Although bound editions will be available for purchase, IJPHM is fundamentally an online journal. As such, we are able to have a very fast turnaround time. We will acknowledge receipt of submissions within three business days, and we intend to rigorously review and return a decision to the authors in approximately 8-12 weeks. Thus, papers may be published in a very short time, allowing your research to be available to the scientific community when it is most relevant.

Option to Present Your Work at a Conference

PHM Society publications have maintained high quality standards for both its Conferences and the Journal. Highest quality conference papers are also invited to be published in the Journal. However, since 2012 IJPHM provides an option to the journal authors to present their journal paper at one of the upcoming PHM conferences.

Authors are reminded that the paper must be journal quality and adhering to the journal template. The paper will be reviewed as per journal review standards and if accepted a presentation slot will be reserved at the target conference. The paper will be published in the journal archives and linked through conference proceedings.

Benefits

- A journal publication of your high quality research work
- A peer review of your work by experts in the field
- A chance to present your work to the targeted audience
- No reworking required to publish in the Journal
- A shortened review cycle to journal publishing

Risks

- Rejected papers will not automatically be considered for the conference and may additionally miss the submission deadline.
- If re-submitted for the conference, they will be reviewed subject to conference review criteria

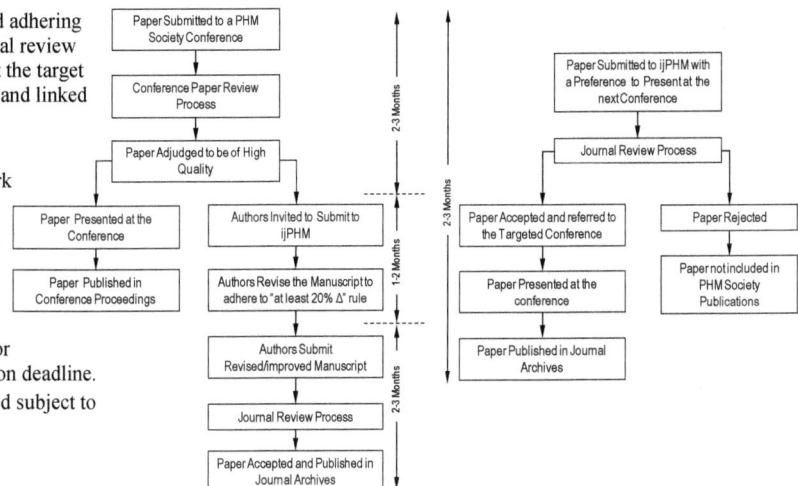

www.ingramcontent.com/pod-product-compliance
Lightning Source LLC
Chambersburg PA
CBHW041450210326
41599CB00004B/203